Wind-Vane Self-Steering

Wind-Vane Self-Steering

How to Plan and Make Your Own

Bill Belcher

Foreword by
H. G. Hasler

 INTERNATIONAL MARINE PUBLISHING COMPANY
CAMDEN, MAINE

Dedicated to Aileen who tries to put up with my enthusiasms

Library of Congress Catalog Card Number 81-85260
International Standard Book Number 0-87742-158-7
Published simultaneously in Great Britain
by David & Charles (Publishers) Limited

Published by International Marine Publishing Company
21 Elm Street, Camden, Maine 04843
(207) 236-4342

Printed in Great Britain

Contents

Part Two: Design and Construction

Foreword

The art of making a sailing boat steer herself by means of a wind-vane gear is relatively young, the first commercial gears having been offered to the public in 1962. Since then a good deal of published material has appeared on the subject, some of it valuable and some of it misleading. I can recommend this book as being thoroughly sound and practical.

Bill Belcher is a man after my own heart: an offshore seaman with a marked talent for empirical research and simple engineering. He has spent a lot of time planning, building and testing vane gears, often on his own boats, and I have gleaned some good ideas from him, notably from the experimental work on vanes and rudders.

There have been other books on vane steering but this compares well with any of them. Theorists may find it less than comprehensive, but it is not supposed to be a text-book of theory. For the reader who wants concise and clear instructions on how to plan and build his own vane-steering gear, this is just about the best book I've yet seen, and I would gladly go to sea with any of the gears he recommends.

H. G. Hasler

Preface

It is obviously impossible when writing a technical book to check personally and verify every statement made. Too often the writer does his research out of other people's books and if, by chance, there was an initial error this then becomes entrenched and the more often it is quoted the more it is accepted as Gospel.

In this book all designs described have been made and used by me. Any comments, apart from those about commercially made gears, are the result of personal experience. For the historical notes in the introduction however reliance has been placed on published sources as it is hard to check those details which, though of interest, do not affect today's designs. For the theory of air and water flow and lift coefficients I have had to rely on published results. These may be in error as the Reynolds number at which the results were obtained is not always stated; but the results are, again, of interest and if not completely accurate do not influence self-steering design.

Where no reliable data was available I used my own primitive wind tunnel for wind-vane investigation, or my boat *Tabitha* as a floating test rig to determine the behaviour of rudders and trim tabs. Complete accuracy cannot be claimed for the results as the methods used lack the refinements necessary for exact measurement, but the data obtained are sufficiently precise to serve as a basis of self-steering assessment and design.

As a child I was advised not to accept my elders' statement that I was being fed a peppermint, but to suck it and see. Excellent advice!

Introduction

When man started to sail, which was basically his method of saying he was tired of rowing, he soon found out that steering a boat was also a fatiguing operation. It cannot have taken him very long to devote his as yet simian brain to the problem of how to get his boat to steer itself. This would allow him to fish, or meditate, or even just lie on the deck and enjoy nature. I have used the word boat here to distinguish it from a ship which had plenty of people on board who could take it in turns to steer. This difference still stands, although the ship now uses electronic gadgets in place of manpower.

The earliest attempts would obviously be made by adjusting the sails and the rudder so that the boat steered itself. This, if the boat was well designed and the right shape with a suitable rig, was satisfactory; but the helm needed constant supervision to avoid going off course when the wind strengthened or fell away. There was also much time involved in getting just that nice balance required.

It was the enthusiastic model-boat sailors on the Round Pond in Kensington Gardens, London, who were the first to see the need for something better than a weighted rudder and a well-balanced sail plan. In 1904 George Braine developed his gear to steer model yachts. This, however, was only an improved method of using the sails to control the steering. The first really effective wind-vane steering was used by S. Berge of Norway, who won the International A Class championship in 1935 with *Prince Charming* fitted with a device controlled by a true wind vane.

Already the benefits conferred on yachting by international cooperation are becoming clear and all development from now on is to be seen to be not the province of one country only, but rather as a relay race where one country has run its length of the track and another has accepted the baton and then passed it on, perhaps to the first runner, perhaps to another, but always in amity and mutual goodwill.

Wind-vane gears for model yachts have not only to hold the boat on course, but have also to be able to tack them if necessary giving long and short legs to the tacks. The Lassel and Fisher gears are, if anything, more complicated than the gears on full-sized yachts where there is at least one man aboard, unlike model-yacht sailing where the skipper, having launched his boat, can only hope that he has all his settings right as he has no chance of altering them until the boat reaches the other end of the pond.

The first well documented use of a self-steering gear on a full-sized boat was in 1936 when the Frenchman Marin Marie used a V-shaped wind vane coupled to his rudder by lines to make an Atlantic crossing in *Arielle*. There was an apparently dead period for the next fifteen years when many of us had other things to do. In 1951 when Pat Ellam and Colin Moodie sailed *Sopranino* across the Atlantic they still relied on a Braine gear where the mainsheet is led to a quadrant on the rudder stock. An elastic line is set with the right tension so that an increase of the force on the mainsheet turns the tiller. If the wind strength decreases, the elastic returns the tiller to its former position (Figs 1 and 2).

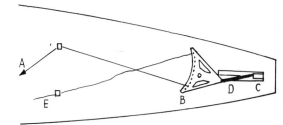

Fig 2 The Braine gear: not a true wind vane as it was worked by the sail pressure via the main sheet. A=mainsheet; B=quadrant on rudder head; C=slide on track; D=elastic; E=sheet for other tack.

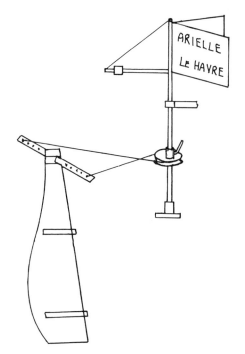

Fig 1 The gear used by Marin Marie on Arielle for an Atlantic crossing in 1936, an early attempt at self-steering. That it worked was probably due to careful sail setting rather than the power of the vane.

In 1955, Ian Major used what appears to have been the first trim-tab controlled rudder, governed by a simple cloth-covered vane, in his successful crossing of the Atlantic in *Buttercup*. In the same year Michael Henderson fitted *Mick the Miller* with a vane working a horn-balanced auxiliary rudder. With 'Harriet', as the gear was called, he won the English J.O.G. Championship, thus emulating the famous greyhound after which the boat was named. John Hetherington, who owned *Mick the Miller* in the 1956 season, found out what anyone who today uses a reasonable self-steering gear knows — the gear is worth an extra hand (on long voyages, two). It requires no sleep, food or water and leaves the human part of the crew free to do chores, sleep, or keep a good look out free of the 'tyranny of the tiller'. In 1957

the famous, if eccentric, French yachtsman Bernard Moitessier fitted trim-tab gears to *Marie Therese* and *Wanda*. A similar gear was fitted in 1964 to Jean Lacombe's *Golif* for the single-handed trans-Atlantic race.

It was the single-handed trans-Atlantic races that really put self-steering on the map. All five competitors in the first race in 1960 used some form of self-steering, and it is not surprising to find the name H. G. Hasler among the competitors. After the race was over he said, 'I've never had to go on deck in an emergency or in bad weather. The wind-vane steering has become part of me, something I shall never sail again without.' He had spent just one hour at the tiller of *Jester* in forty-eight days; his self-steering did the rest. Hasler's designs may have been improved upon, but he certainly set the standard for all other designers to aim at.

In this book all dimensions are given in metric units, although some may seem a little odd. This is due to the gears having been developed at various times over a number of years. Those which were actually designed in inch units have had the inch dimensions converted to the nearest round-off metric equivalent and *vice versa*. Later designs were worked out in metric units and will, therefore, seem more logical.

PART ONE

Theory and General Principles

CHAPTER ONE

Types of Self-steering gear

This book treats only of wind-vane self-steering gears, but brief mention should be made of two other methods of getting a boat to sail a course unattended.

The Autopilot is an electronic means of steering where the direction is sensed by electronic sensors on a compass or small wind vane. This type of gear can be made very sensitive and usually has built-in controls so that the steering can be adjusted, avoiding on the one hand violent oversteer and on the other too slow a response which does not give quick enough recovery to the boat if there is a wind change or a large wave pushes it off course. These built-in controls make the gears adjustable for boats with many different steering characteristics and avoid having to tailor-make the gear to suit the boat if the very best performance is required.

The disadvantage of Autopilots is that, if they fail in service, they are not repairable by the electronically unskilled sailor. Also, and this is important, they require electricity to work them. On larger boats, particularly the motor sailers with all sorts of electrical gadgets and almost certainly a spare generator on board, this is no problem; but on small yachts, navigational lights give quite enough drain on the batteries without adding an Autopilot. Under favourable conditions some of the latter have only a low-current consumption, but when the going gets tough will certainly draw more current than the makers, hopefully, claim.

A further disadvantage is that, in certain Autopilots, in order to save the mechanism from overload burn-out, there is an overload disengagement. This is fine and protects the Autopilot, but I found that under hard-pressed storm conditions, when the Autopilot could cut out at any moment, things could get very dicey in a very short space of time.

The other method of self-steering is by the time honoured method of so adjusting the sails that the boat sails itself on the course desired with the tiller lashed or controlled by lines from the sails. This method has the disadvantage that it is usually slow to set up, requires frequent adjustment with rising and falling wind strengths, and very often the sails are used not to best advantage to drive the boat forward but so that, by adjusting one against the other, the boat can be sailed in the desired direction. This art, and art it certainly is, should however be practised by every yachtsman who is seriously considering ocean voyaging. According to Murphy's Law, 'If anything can go wrong, it will'. If as well as your self-steering gear you can, at a pinch, use your sails to steer the boat, you may yet fool that troublesome Irishman.

Autopilots have one big advantage over wind-vane gears — they are controlled by a compass or wind vane whose position is sensed in one of a number of different ways, magnetic, optical, or electronic. This does not require any work from the wind vane which is thus free to face directly into the wind at all times. This means that the course set at a predetermined angle to the relative wind can be held with considerable accuracy. With

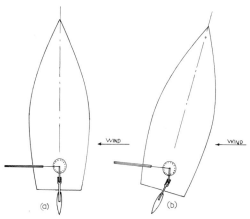

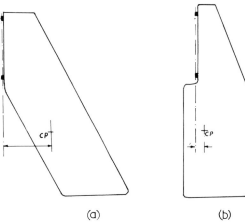

Fig 3 *Wind-vane steering: (a) cruising; (b) as the wind freshens the boat tends to luff to windward. This goes on until there is enough pressure on the vane to turn the rudder, after which the boat will stabilise on the new course.*

Fig 4 *Rudder (a) is hard to turn, but rudder (b) which is balanced by having the leading edge well forward of the pintle axis can be turned with ease. CP (the centre of pressure) is the point on the rudder at which the forces can be said to act.*

wind-vane steering, the vane itself, either directly or through some means of amplifying the force on the wind vane, works the rudder. This requires some force equivalent to the helmsman correcting the boat's tendency to go off course by pulling or pushing the tiller. But the wind vane can only generate a force by being at an angle to the wind; and the greater the angle of the vane to the wind, the greater the force. This is the snag.

We want to hold a reasonable course to the relative wind, that is to say, to the wind we feel when we stand on the deck. Variation in the wind strength will mean that more or less force on the rudder is required and this can only be effected by more or less force on the vane. If the vane is large a very small change of angle to the relative wind will produce enough force for the rudder correction and this small angle can be tolerated as a course variation (Fig 3). If however the vane is not of sufficient size, the boat will luff up to an unacceptable degree before there is enough force for the wind to control the rudder. In practice, as the wind strength increases there will be more force on the wind vane for the same angle of deflection. The increase of force on the vane,

however, is not enough in most cases to cope with the extra force required on the rudder. The result is that boats under self-steering tend to luff up to windward as the wind strengthens, and bear away as it dies.

There are several ways out of this difficulty, and they are relative and not absolute: either use a large or powerful wind vane or a rudder which, because of its design, requires very little force to turn it; or in some way increase the force of the vane before it is applied to the rudder (Fig 4). The art of self-steering design is to orchestrate these separate elements so that we end up with a gear which has full control of the boat without having to have the vane at an unacceptable angle to the relative wind.

Before going further, we should consider this relative wind and how it affects us. If we are beating into wind, the speed of the wind across the deck is a vector combination of the boat's speed plus the speed of the wind. If we are running — that is running away from the wind — the speed over the deck, which is what works our wind vane, is the speed of the wind minus our forward speed.

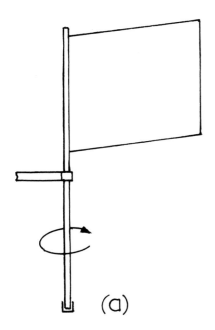
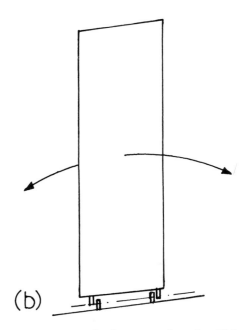

Fig 5 Vanes can be either vertically pivoted as at (a) or horizontally pivoted as at (b).

To windward any reasonably designed boat will probably sail itself, but running is an altogether different matter. If our boat is small, which due to financial stringency mine usually is, with a following wind of, say, 12 knots we will perhaps be making 4 knots, giving us an effective 8 knots over the deck. This is enough to work any reasonable gear. If however our vessel is larger or swifter, such as a multi-hull, we will be doing perhaps 7 knots with the same wind. We now have only 5 knots to work our gear. As the force on the wind vane is proportional to the square of the wind speed, the effect on the wind vane is not in the ratio of 8:5, but of 64:25; that is to say, on a bigger boat we have only 40 per cent of our power. The moral is that, the bigger the boat, the harder it is to steer on downwind courses.

It is not only the size, but the shape of the boat which dictates what is required. My last boat, the unfortunate *Josephine*, was of rather unusual shape as it was built, among other things, to be easy to self-steer. She could be sailed with a foot of water over the lee gunnel and still be held on a straight course with one finger on the tiller. It is true that at this angle the helmsman had one hand and arm fully employed trying to stay in the boat, but the actual steering was no difficulty. My present vessel, *Tabitha*, designed — not by me — for families and the 'lean and slippered pantaloon', has lots of space; but if the water gets anywhere near the lee gunnel she just heads up to windward whatever the rudder is doing to check this motion. You have to let the mainsheet go just as if the boat were an over-canvassed dinghy. *Tabitha* will self-steer, but only by keeping the sail plan well forward and the boat as upright as possible. *Josephine* might throw you out of the cockpit or your bunk, but would keep on sailing more or less where you wanted to go. She now rests on the Middleton Reef — but that's another story.

The lesson to be learnt from this is that each boat should be judged on its own merits and the capability of its owner. It is possible to make a self-steering gear so powerful that it will override any defect of hull shape or the owner's knowledge of sail setting, but it will certainly be heavy

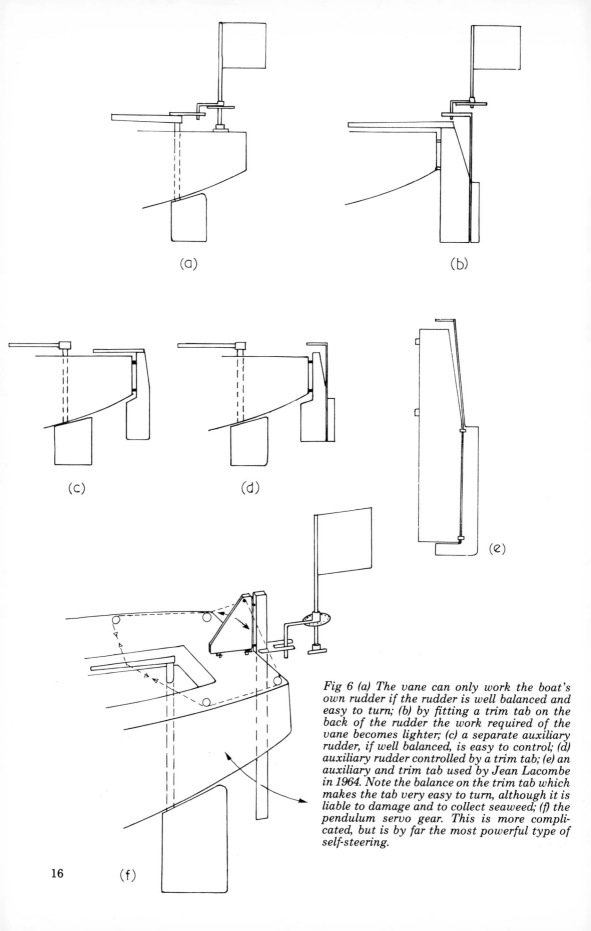

(a)

(b)

(c)

(d)

(e)

Fig 6 (a) The vane can only work the boat's own rudder if the rudder is well balanced and easy to turn; (b) by fitting a trim tab on the back of the rudder the work required of the vane becomes lighter; (c) a separate auxiliary rudder, if well balanced, is easy to control; (d) auxiliary rudder controlled by a trim tab; (e) an auxiliary and trim tab used by Jean Lacombe in 1964. Note the balance on the trim tab which makes the tab very easy to turn, although it is liable to damage and to collect seaweed; (f) the pendulum servo gear. This is more complicated, but is by far the most powerful type of self-steering.

16   (f)

and probably expensive. On the other hand, a light, cheap gear in the hands of someone who knows what he is doing and knows how to get the best out of his gear will very often serve as well, or nearly as well, as a much more expensive one. I must confess here that the first self-steering gear I purchased was given away to someone in the marina at Gibraltar. Subsequent interest and experiment with gears has confirmed me in the opinion that it was not the gear that was at fault, but rather my ignorance of how to work it successfully. By careful study of the various types of gear available the reader will be able to judge the relative merits of the different systems, and opt for the one which best suits the work required of it — and his pocket book.

The question of work required must not be neglected. Is the gear to be an assistant to the helmsman, giving him a chance to make a cup of tea, get his fenders out at the end of the trip, do a bit of navigation, or any one of the countless tasks a single-handed sailor cannot do when he is tied to his tiller? Or is the boat owner a 'press on' type who wants to enter single-handed races and needs a gear which will steer his boat through a gale whilst he is nice and warm asleep in his bunk?

Before going into the finer points of design, it is well to give a rough idea of the self-steering methods presently used. Wind vanes are either vertically pivoted (VP) vanes, or horizontally pivoted (HP)

vanes. The vertically pivoted ones are like stiff flags on a vertical shaft, whereas the horizontally pivoted are hinged about their lower end and are blown out of the vertical by the wind (Fig 5). These vanes work some means of steering a boat which can be (Fig 6):

The vane working the boat's rudder directly.

The vane working a trim tab attached to the rudder to get the necessary power from the wind vane to turn the rudder.

The vane working an auxiliary rudder which is a separate rudder hung on the boat's stern and may be unbalanced, balanced, or worked by a trim tab.

The pendulum servo system.

The pendulum servo system is not so easy to understand as the others, but consists of a balanced servo blade which is in the water. This blade is long and narrow and as its vertical axis is near the centre of pressure it is very easily turned and, as soon as it is turned, at an angle to the water flow, a large sideways force is generated. This swings the blade over like a pendulum. Lines attached to the pendulum are led to the tiller or wheel.

There are many interesting hybrids of these basic gears. One of these, the Saye gear from America, is part pendulum servo, part trim tab. It is more fully described on page 79.

CHAPTER TWO

Theoretical Considerations

This chapter is inserted here as it is on theory that successful design of both wind vanes in the air and rudders in the water depends. The many readers who only want practical hints and ideas and are quite definitely turned off by theoretical considerations can quite easily skip this chaper and get on with the practical side; but the theory will give those of scientific bent a basis for evaluating the reasons for statements made later on.

Lift and Drag

It is at first a little difficult to understand but, broadly speaking, any body, whether immersed in a moving liquid such as water, or a gas like air, behaves in the same manner. Provided we know the shape of the body, the density or weight per unit volume of the gas or liquid and the speed at which the body is travelling through the medium, we can calculate the forces on it.

When a flat plate is put edge-on to a fluid stream there is a certain small amount of drag in the direction of the flow, but no other force (Fig 7a). But, as b, c and d show, when we incline the plate at a small angle, the way that particles move past the plate, streamlines as they are called, is altered. We now have a force on the plate acting at right angles to the fluid flow, which we call the lift, and a slightly increased resistance, or drag.

The lift on the plate is caused by the fluid hitting the underside of the plate. This is easy to understand. What is not so easy to see is that the curved path of the particles on the top of the plate creates suction on the topside, giving increased lift. When we incline our plate even further we still get a pressure on the underside caused by having a bigger area exposed to the flow, but the flow behind the top edge has broken down and instead of getting streamlines, or even flow, we get turbulence. This gives us no lift, but increases the drag. Finally, there is a complete breakdown in the flow past the plate giving us a lot of drag, but little lift.

It will be seen that the loss of lift was created by the turbulence in the fluid which, beyond a small angle, was not able to follow round the sharp forward edge of the plate. If now we use a shape which is not a flat plate, but is of 'streamline' section, we can expect to get greater lift because we will not get the early formation of turbulence (Fig 8a).

The streamline shape has the same pressure on the underside as a flat plate, but the fluid is able to follow round the nose of the foil and thus we get a far greater suction on the topside. When, however, we incline the shape at an angle which is too great, the fluid no longer sticks to the topside, but separates (Fig 8b).

This occurs at about 15° to 18° and results in loss of lift and considerable increase of drag. The angle at which this breakdown occurs is known at the stalling angle.

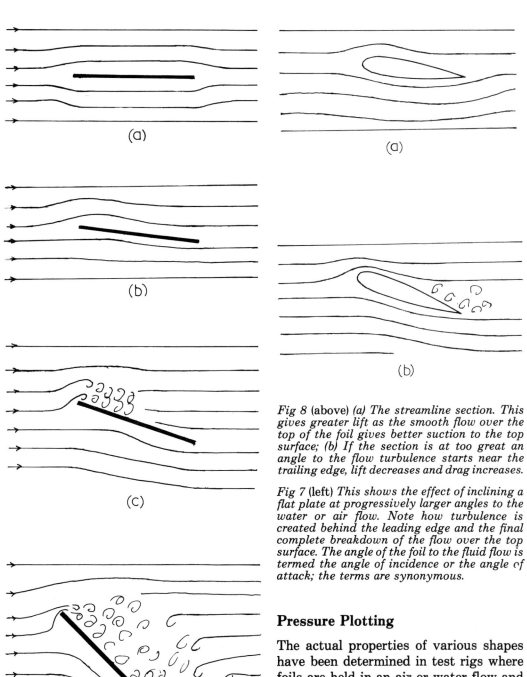

(a)

(b)

(c)

(d)

(a)

(b)

Fig 8 (above) *(a) The streamline section. This gives greater lift as the smooth flow over the top of the foil gives better suction to the top surface; (b) If the section is at too great an angle to the flow turbulence starts near the trailing edge, lift decreases and drag increases.*

Fig 7 (left) *This shows the effect of inclining a flat plate at progressively larger angles to the water or air flow. Note how turbulence is created behind the leading edge and the final complete breakdown of the flow over the top surface. The angle of the foil to the fluid flow is termed the angle of incidence or the angle of attack; the terms are synonymous.*

Pressure Plotting

The actual properties of various shapes have been determined in test rigs where foils are held in an air or water flow and the forces acting on them calibrated. Use is also made of 'pressure plotting', for which a streamline shape has a large number of holes in the surface. Internally, these holes are connected to manometers, or pressure gauges, and give the distribution of suction or pressure over the

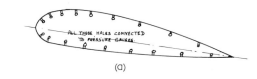

(a)

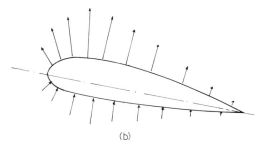

(b)

Fig 9 (a) Pressure plotting gives an accurate measurement of the forces acting on a foil; (b) The arrows indicate direction and magnitude of the forces. Note the suction on the top of the foil is greater than the pressure underneath and how the front end of the foil does most of the work.

surface (Fig 9a). The results of these tests give pressures normal, or at right angles, to the surface. The length of the arrows in Fig 9b indicates the size of the force. It will be seen that the suction on the topside is considerably greater than the pressure on the lower side and also that the suction and pressure are not uniform over the surface but are concentrated at the forward part of the foil which thus does most of the work.

For a flat plate the resultant pressure acts at about 33 per cent of the chord (the chord is the width of the plate or foil) back from the leading edge and tends to move further aft with increasing angle of the foil to the airstream. With a streamline section the centre of pressure is further forward at 24 to 29 per cent of the chord depending on the streamline shape used.

For our rudders, we want our section to be symmetrical, as it is no good having a shape like an airplane's wing with one side flattish and the other side curved. This works fine one way, but is not much

use the other. The section should be reasonably strong without a very fine trailing edge which is easily damaged. Finally, it will help if the centre of pressure on the foil is predictable and fairly constant in position.

American NACA 00 Series

The American NACA 00 series has all the properties we want and is the section we will adopt for all our examples. The figure 00 indicates that the section is symmetrical. A further two numbers indicate the percentage ratio of the thickness to the chord. The section 0012, for instance, is a symmetrical section of a thickness which is 12 per cent of the chord. In all cases with this series the maximum thickness is at 30 per cent aft of the leading edge. The centre of pressure, which is the resultant of all forces acting on the foil, is practically constant at 24 per cent of the chord from the leading edge. It is only when the foil stalls and the streamline flow breaks down that the centre of pressure moves aft to 30 per cent or more. In actual practice, due to roughness, or not quite perfect shaping, the centre of pressure can be taken at about 25 per cent aft of the leading edge.

The table gives the coordinates of the NACA 0010 section; this has a thickness of 10 per cent of the chord. For any other thickness to chord ratio it is only necessary to multiply the offsets to correspond to the new ratio. If we want a rudder of 400mm chord with a thickness, for strength reasons, of 75mm, then the thickness ratio will be 75÷400, or 18.75 per cent. This is a little thick as 15 per cent is about right, but anywhere between 10 and 20 per cent can be used.

To obtain the required offsets, that is to say, the distance from the centreline of the section to the surface, it is only necessary to multiply the tabulated offsets by the new ratio. For example, in the table, the offset for 20 per cent of the

NACA 0010

Percentage distance from leading edge	Offset to upper and lower surface Percentage of chord
0	0
1.25	1.57
2.5	2.17
5	2.97
7.5	3.5
10	3.9
15	4.45
20	4.78
30	5
40	4.83
50	4.42
60	3.83
70	3.06
80	2.18
90	1.21
95	.67
100	.02

chord is given as 4.78 per cent. The new percentage will be

$$\frac{4.78}{10} \times 18.75$$

As this figure is still only a percentage of the chord, the actual value for our section with a chord of 400mm will be

$$\frac{4.78 \times 18.75}{10} \times \frac{400}{100} = 35.8\text{mm}$$

The offsets to complete the section can be calculated the same way.

Forces on a Foil

When a foil is placed at an angle in a moving stream of fluid the magnitude of the two forces developed — at right angles to the fluid flow, the lift; and in the direction of the flow, the drag — is determined by the density and velocity of the fluid hitting the foil and the actual shape of the foil. The effect of the foil shape and angle of attack as determined by experiment, gives coefficients which are dimensionless and do not change whatever system of units we use. The pressure derived

from the density and velocity is called the dynamic pressure and is expressed as

$$\text{Dynamic pressure} = \frac{dV^2}{2g} = q$$

where d = density in kg per m^3
V = velocity in m per sec
g = 9.81 m per sec^2

The dynamic pressure for air

$= 0.0166V^2$ kg per m^2 where V = wind speed in knots

When water is the fluid, dynamic pressure

$= 13.58V^2$ kg per m^2 where V = speed in knots in fresh water
$= 13.91V^2$ kg per m^2 where V = speed in knots in salt water

The actual force acting on the foil is

L (or D) = C_L (or C_D) $\times$ q $\times$ S
where L = total lift or force at right angles to the fluid flow
D = total drag or force in the direction of the fluid flow
C_L, C_D = coefficients of lift or drag determined by experiment
q = dynamic pressure
S = area of the foil

This equation only works if all units used for dynamic pressure, area and lift are in the same units, either ft lb or metric. The coefficients of lift and drag are the same whatever units are used.

The coefficients of lift and drag corresponding to the foil shape have been determined from experiment and, although they do vary with the density and velocity of the fluid, are reasonably accurate for practical purposes.

There is, however, one respect in which the coefficients can be seriously in error: that is when the aspect ratio of the foil is considered. The aspect ratio is, in its simplest form, the ratio of the length of a foil to its width, or chord (Figs 10, 11). When, however, the foil is not of uniform

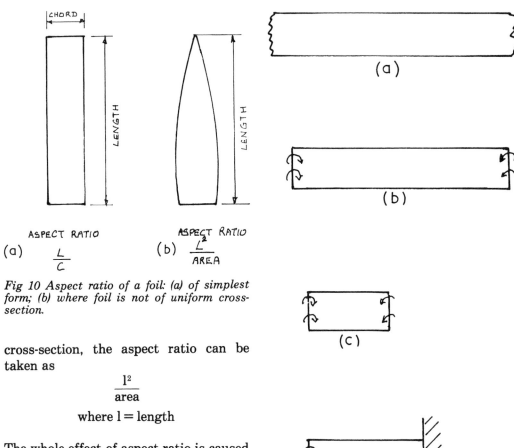

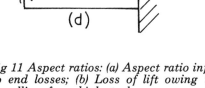

Fig 10 Aspect ratio of a foil: (a) of simplest form; (b) where foil is not of uniform cross-section.

cross-section, the aspect ratio can be taken as

$$\frac{l^2}{area}$$

where l = length

The whole effect of aspect ratio is caused by the fluid flowing from the high pressure side to the low pressure side, round the end of the foil. A foil of unlimited length would have no end losses. As, however, the aspect ratio is reduced, the loss becomes more and more considerable. If we blank off one end of our foil then there is no loss at one end, so the effect is to double the effective aspect ratio. With a boat's rudder, if the rudder is under the hull with a minimum gap between the rudder and hull, the effective aspect ratio has been determined as 1.7 times the ratio derived from the geometry of the rudder. Deck-sweeping sails also derive their efficiency by not allowing air to escape from the high to the low pressure side underneath the sail.

Fig 12 gives the coefficient of lift for a curved section of various aspect ratios and also for a flat plate of AR5 when the

Fig 11 Aspect ratios: (a) Aspect ratio infinity, no end losses; (b) Loss of lift owing to air travelling from high- to low-pressure side of foil; (c) With low-aspect ratio the proportion of loss to total lift increases; (d) This foil only loses efficiency at one end, and has the same coefficient of lift as a foil of nearly twice the aspect ratio.

foils are in the water. It will be seen from this that the high-aspect ratio foils give a greater maximum lift, but stall at a lower angle of attack than a foil of lower aspect ratio. The flat plate, although initially about as good as a shaped section, falls off badly after a deflection of about 8°.

Fig 13 gives the same coefficients as Fig 12, but with the foils in air. It will be

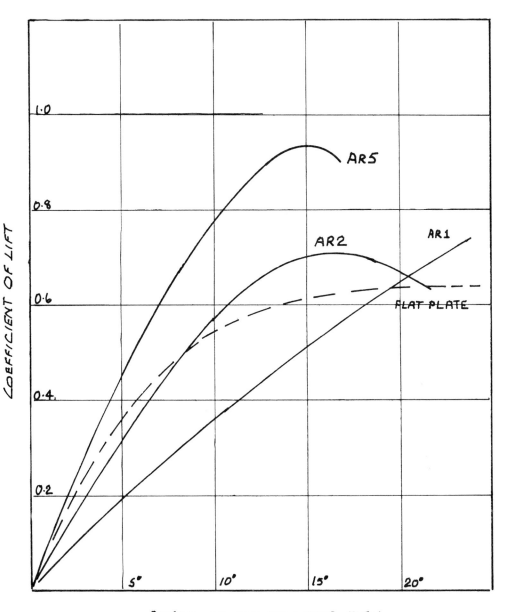

Fig 12 The coefficients of lift for foils in water show clearly the effect of aspect ratio and the superiority of the streamline foil over the flat plate at anything but small angles of incidence.

23

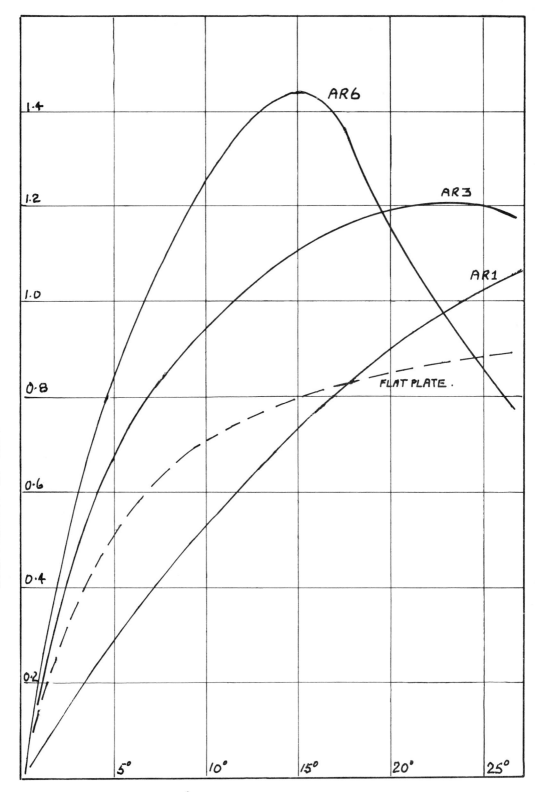

ANGLE OF FOIL TO AIRFLOW.

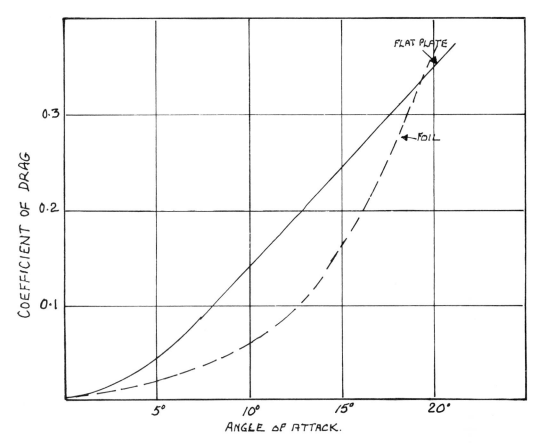

Fig 14 An illustration of how the drag of a streamline section is considerably lower than that of a flat plate until the stalling angle is reached.

seen that, although the coefficients have not exactly the same values, the general picture is the same.

Fig 14 shows the difference between the drag on a flat plate and a shaped foil. It will be seen that over normal angles of attack the flat plate has considerably more drag than the shaped foil. It is only when the shaped foil has stalled that its drag becomes the same as the drag of a flat plate.

Fig 13 Coefficients of lift for foils in air: although these differ from the coefficients obtained for foils in water owing to the very great difference in density between air and water, the picture is still the same as that for water.

In the actual field of operations for self-steering gears, we use flat plates for wind vanes, or flat plates with 'ears' on the trailing edge. The reason for this is that we want the centre of pressure of the wind on the vane to be as far back as possible and a flat plate has a centre of pressure at about 33 per cent aft of the leading edge, whereas the shaped foil has the centre of pressure at 25 per cent. It is true that the coefficient of lift is lower than that for a streamline section, but this can be compensated for by a slight increase in area. For small angles, however, there is little difference, and the lift obtained from a flat or curved foil is very similar up to about 8° of deflection of the vane. As we are trying to make a gear which, hopefully, does not wander far off course, a vane which has its efficient range up to 8° is quite good enough.

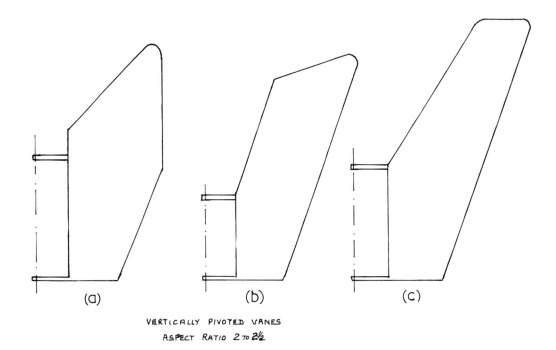

VERTICALLY PIVOTED VANES
ASPECT RATIO 2 to 2½

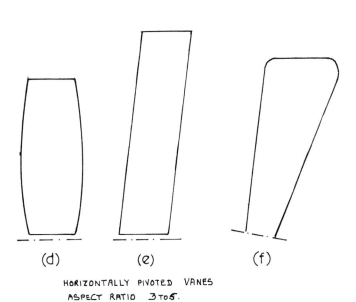

HORIZONTALLY PIVOTED VANES
ASPECT RATIO 3 to 5.

Fig 15 (a), (b), (c) Different shapes of commercial vertical-axis vanes; the aspect ratio is from 2 to 2.5. Vane (a) is my 'trade mark'; (d), (e), (f) Shapes of horizontal-axis vanes currently in use.

In practice, the aspect ratio of the vanes and rudders we use usually lies within fairly closely defined limits due to structural and space limitation.

Vertical-axis wind-
   vane aspect ratio   2 to 2½
                             (Fig 15a, b, c)
Horizontal-axis
   wind-vane aspect   3 to 5
   ratio   (Fig 15d, e, f)
Balanced rudders   2 to 2.5 (Fig 16)
Pendulum servo
   blades   3.5 to 4.5 (Fig 17).

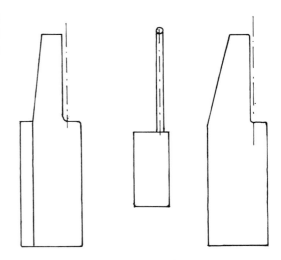

Fig 16 Auxiliary rudders have an aspect ratio of 2 to 2.5 chiefly governed by strength considerations; they should also be well balanced.

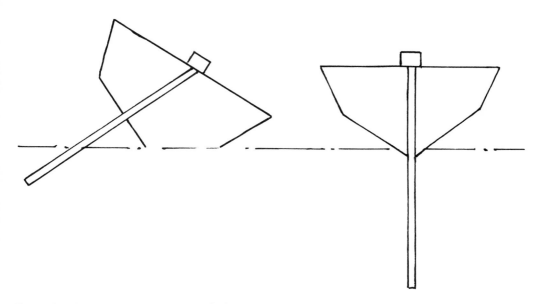

Fig 17 Pendulum servo blades must be long to keep in the water, and their small chord gives an aspect ratio of 3.5 to 4.5.

27

CHAPTER THREE

Wind Vanes

Any vane steering gear must have a vane to provide the initial force used to effect the steering. This vane is connected by a linkage either through a servo system or directly to a rudder, which may be either the boat's rudder or a separate auxiliary rudder. This linkage can be quite a complicated set of levers and gears or, in its simplest form, a piece of string. It is proposed in Part One to treat each of these parts separately, ie wind vanes, linkages, rudders and servo systems. In Part Two these will be built together into practical working gears which can be used by anyone who, rather than buying expensive equipment, would like to understand and build his own.

Like nearly all parts of self-steering gears, wind vanes are largely a question of personal opinion. They also, by their distinctive shape, serve as the maker's trade mark, rather like the heraldic shields of our ancestors. There are basically two schools of thought. One says that the vane must be as small as possible, as any wind vane is a blot on the fair lines of their boat. The other, more utilitarian, takes no notice of the look of the gear, but wants one that works as well as is humanly possible to make it.

The first wind vanes were plain, if rather large, vanes mounted on a vertical axis and coupled directly to the rudder of the boat. Sir Francis Chichester used one like this which he called 'Miranda'. She was a lady of very ample proportions, more like a mizzen sail than a wind vane, but apparently she worked. Several devices were tried to increase the efficiency and reduce the vane sizes, but before considering these perhaps a little theory will set our feet on the right path.

A vane should be as light as possible, as weight means friction and we want our vanes to respond to the lightest of airs. It should be of a shape which gives good lift at small angles of attack, and the centre of pressure should be as far back from the pivot as possible.

If we look at our curves of lift coefficients (Figs 12, 13) we see that the lift generated by a flat plate is similar to that of a streamline section for small angles of attack. It is only when the angle of attack is more than about 8° that the streamline section really comes into its own. Thus, if the wind vane is to work at only small angles of attack, then a flat plate of slightly larger size than a streamline section is sufficient. The flat plate also has the practical advantage that it can be made of a piece of plywood which is probably lighter and certainly easier to make than the equivalent shaped section.

The major advantage of the flat plate is in the position of the centre of pressure, the point at which the resultant of all forces acting in the vane acts. The fact that the flat plate's centre of pressure lies at about 33 per cent of the chord aft of the leading edge whereas the streamline section's lies further forward at 24 to 28 per cent dependent on the section used, means that the leverage the flat plate exerts about its forward edge is greater than that of the streamline section (Fig 18). This benefit only applies to vertical-axis vanes.

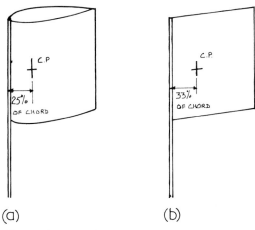

(a)   (b)

Fig 18 As a wind vane, the streamline section (a) has a disadvantage in that it is probably heavy and, although the lift generated is good, the centre of pressure is too far forward to give good torque. The flat plate (b) gives about the same torque as a streamline section, but is much lighter and easier to make.

The turning effect of the vane is proportional to the area multiplied by the distance of the centre of pressure from the axis of the vane. Without increasing the dimensions of the vane we can increase its effectiveness by moving the vane away from the vane pivot, or we can cause the centre of pressure to move aft by fitting 'flaps' on the after edge of the vane. These flaps have a double benefit: they move the

Fig 19 Types of vertical-axis vanes: (a) V-shaped vane; (b) plain vane set out from axis; (c) plain vane with Y-shaped trailing-edge flaps; (d) plain vane with T-shaped trailing-edge flaps

centre of pressure to a position near 40 per cent of the chord; they also give extra lift to the vane.

Vertically Pivoted Vanes

Optimum Shape

There are many variations of shape, those in use being commonly variations on one of the types illustrated in Figs 18 and 19. It seemed to be necessary, however, to find out which of these shapes was the best and to try to determine why. A test rig was constructed in which a standard vane, a flat plate, could be compared with shaped vanes (Fig 20). This rig would be placed in the open air where there was a consistent wind and the relative forces generated by the wind in the vanes could be determined. Measurement was only a question of placing a pin in the appropriate hole so that, under the influence of the wind, the two vanes gave equal turning force and neither overpowered the other. The force generated by the vanes would then be inversely proportional to their lever arms (l_1, l_2). Unfortunately, like many other brilliant ideas, the rig did not work with anything like reasonable accuracy as the wind could not be relied upon for either constant strength or direction and, unless some other method of greater precision could be found, the series of tests would have to be abandoned.

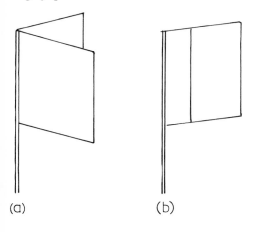
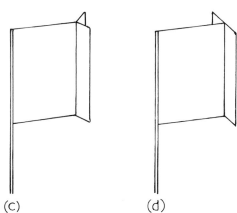

(a)   (b)   (c)   (d)

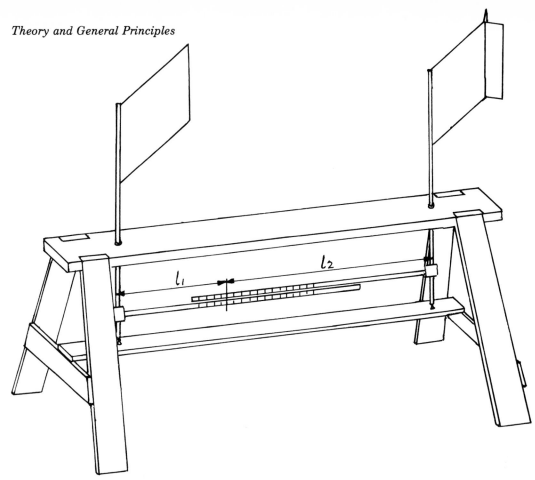

Fig 20 This rig with which it was hoped to measure the relative efficiencies of wind vanes gave results which were inconclusive as the wind velocity and direction were too unreliable.

Fig 21 A wind tunnel set up to test wind-vane torque.

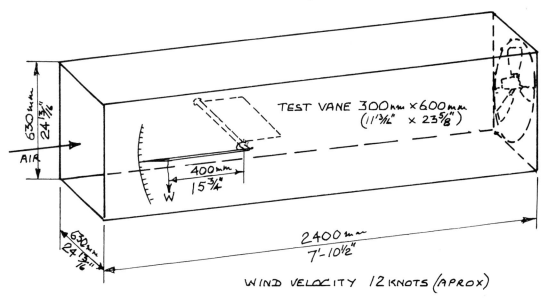

It now seemed that the only thing to do was to construct a wind tunnel. And although limitations of space and finance resulted in a tunnel which was probably too small to give accurate results, it did at least give results which could be checked and duplicated (Fig 21). Under test, the vane was balanced with the fan off. The fan was then switched on and the angle at which the vane lay measured. Weights were then added (at W) until the vanes were approximately 14° from the horizontal. Beyond this there would obviously be false results due to the size of the tunnel which would be partly blocked with the vane acting as a flue damper. Anyway it would be near the stalling angle of the vane, and if the wind were at greater angles than these it would result in very sloppy self-steering. The weights were then reduced until the vane lay at the same angle to the horizontal on the opposite side. The readings were then averaged to eliminate any errors. Fig 22 shows the wind tunnel rigged up to measure the drag on the wind vanes.

The wind vanes tested in our tunnel were all of the same overall dimensions, ie 600mm wide by 300mm chord. There were three flat blades of 175, 225 and 300mm chord with corresponding gaps between vane and shaft (Fig 23). The

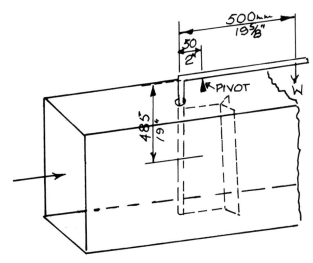

Fig 22 The same wind tunnel set up to determine drag at zero angle of attack.

'results of the tests (on the flat vanes) which did not agree with simple theory, are given in Fig 24 where it will be seen that cutting away part of the vane has little effect as the loss in area is compensated by the increased distance from the pivot to the centre of pressure.

Subsequently the flat plate without cut-out was fitted with various shapes of flaps and two V-shaped vanes were also tried (Fig 25).

In order to avoid an infinity of curves, which only serve to show how diligent the

Fig 23 Cutting out part of a flat vane next to the pivot reduces the vane area, but increases the lever arm.

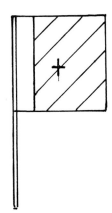
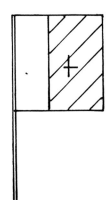
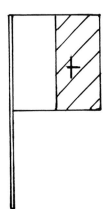

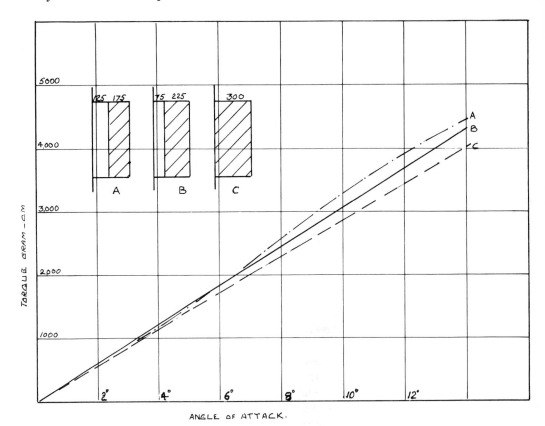

Fig 24 The results of tests in the wind tunnel: these do not conform to theory; this was probably due to the comparatively thick vane shaft.

Fig 25 Different configurations tested in the wind tunnel.

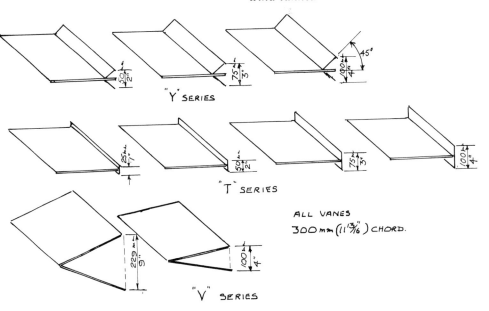

"Y" SERIES

"T" SERIES

ALL VANES
300 mm (11 3/16") CHORD.

"V" SERIES

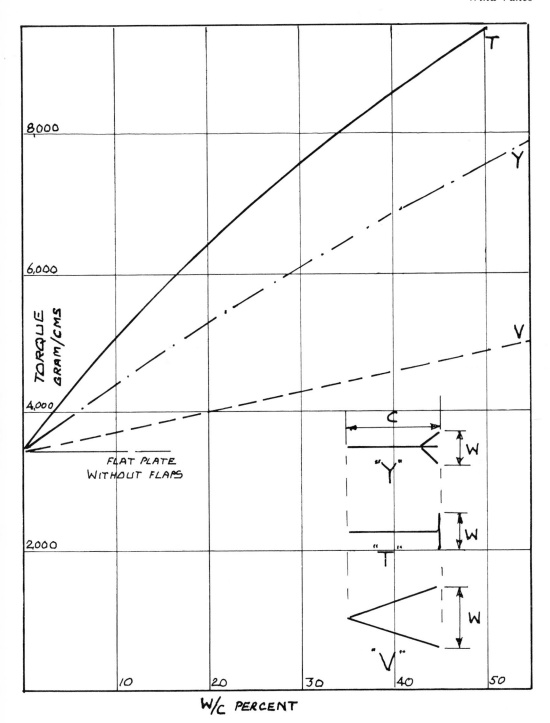

Fig 26 Torque generated by different configurations of vane at 10° to the air flow.

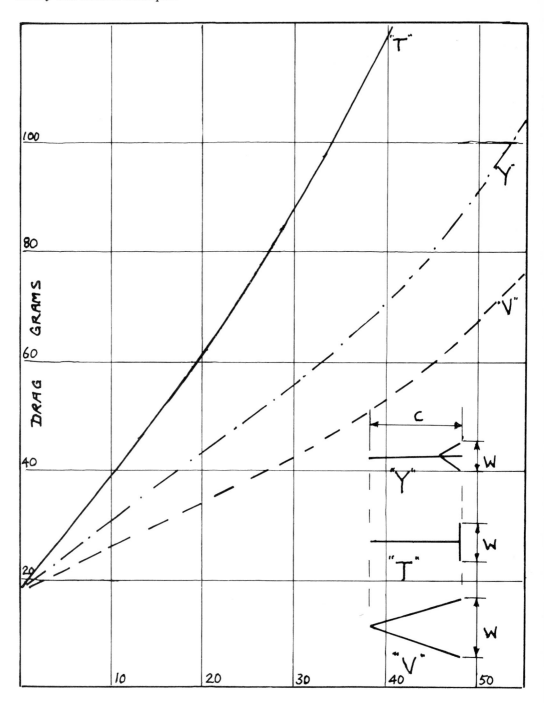

Fig 27 *The drag of vanes at zero angle of*
attack.

observer was and usually result in mental indigestion for the reader, only one set of results has been given (Fig 26). These show the torque given by a vane without cut out, fitted with different sizes and types of trailing-edge flaps when at an angle of 10° to the air flow. There are also results for the V-shaped vane. All vanes had the same overall chord. This gives an easily appreciated picture and would seem to show that the vane with T-shaped flaps gives the best results. There is only one snag and this is the drag.

Fig 27 gives the drag for the various configurations when the vane is facing directly into wind. Here it will be seen why the trailing-edge flaps are not quite such a good idea. The T-shaped vane has perhaps 2½ times the torque, but at least 6 times the drag for the same chord size. This drag means that the wind force on the vane is much greater; thus a heavier shaft will be required and we will get more friction on our bearings.

Provided we have the space it is better to design a slightly larger flat vane. It is not so very much larger as the torque produced is proportional to the chord multiplied by the lever arm or, effectively, to the square of the chord. To obtain the same results as a 50 per cent T-shape we must increase the chord of the flat plate by between 50 and 60 per cent. The interesting point here is that, assuming we are using plywood, we use no more material with a large flat vane than we would if we used a smaller vane with ears. The use of ears should only be resorted to where extra power is required and no space is available, and then only after checking that the shaft is strong enough or else there is a chance that the whole lot may end up overboard.

The actual overall shape of the vane is not of great importance. It should have as much leverage as possible, it should not hit the backstay and it should be as high off the deck as strength permits, as the higher up the better the wind and the less

chance there is of interference by helmsman or passengers standing up in the cockpit.

My standard vane which is used on both pendulum servo and trim-tab gears is of the shape shown in Fig 28. There is a perfectly logical reason for this shape. Four vanes can be cut from a standard sheet of plywood (Fig 29). The bits cut off the bottom of the vane are to miss a backstay with the vane mounted outboard. This can be fitted with ears to increase the turning force (Fig 30).

When comparing the efficiency of different vane configurations care must be taken that the comparison is a true one. Claims of a fantastic break through were made for an American configuration which used a streamline shape with carefully constructed ears set at a specified angle, with a small gap between the ear and the vane. This was stated to produce something like 12 times the torque of a plain vane. Attempts to reproduce the results were a dismal failure until it was noticed that the flat plate, against which the new vane was compared, was not pivoted away from the plate, but the pivot was actually on the surface of the plate. (Fig 31). This meant that the flat plate was operating very inefficiently, as the lever arm from the centre of pressure to the pivot was very short; but the flaps added to the vane moved the centre of pressure aft, increasing the lever arm considerably, and produced the much better performance. Moving the vane axis away from the vane would achieve something like the same effect (Fig 30). Allow for the extra factor of about 1.8 for the actual presence of the ears and the results are credible if misleading.

It will be noted that I did not test the streamline shape. This was for the following reasons. In practice it is hard to make a streamline piece which is the right shape, is light, and is not subject to damage when stowed. Also, the turning

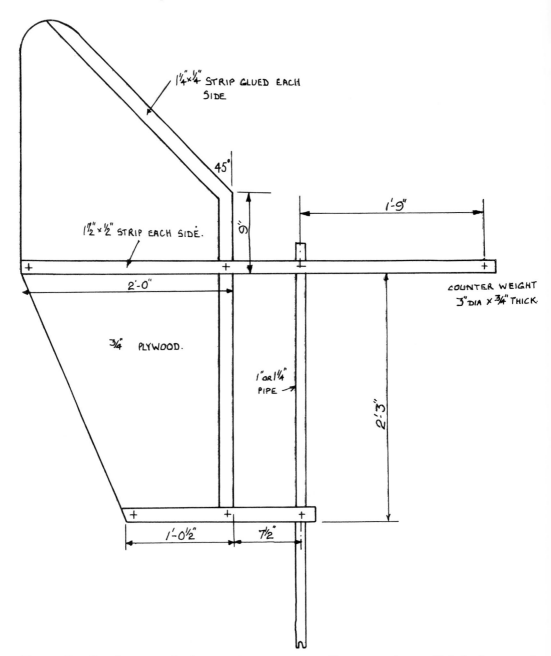

Fig 28 Details of my standard vertical-axis wind vane.

power of a vane is based on the area of the vane and the distance of the centre of pressure from the pivot. With the streamline shape the centre of pressure is only 25 per cent of the chord back from the leading edge so that, although the streamline may give a slightly increased lift, it loses out in the length of the lever arm.

It is interesting in this respect to calculate in theory what happens with a plain, flat vane with the centre of pressure 33 per cent from the leading edge when we cut out various proportions of the vane next to the pivot. If the vane is

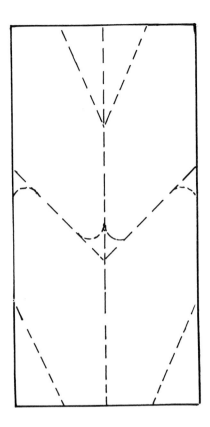

Fig 29 How four vanes can be cut from a standard sheet of plywood.

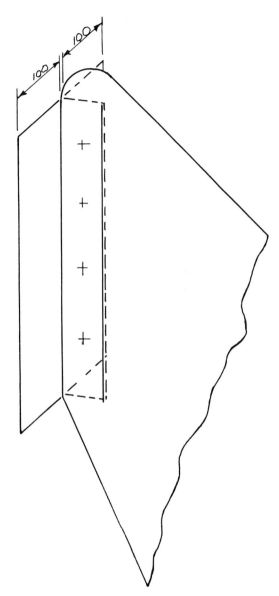

Fig 30 Ears added to a vertical-axis vane considerably increase the torque, but at the expense of extra drag.

plain and flat and the pivot of negligible thickness the best vane is, theoretically, one that has a cut-out equal to 25 per cent of the overall chord. In practice however it was found (see Fig 24) that this was not so, although there was little difference between the different vanes. The difference is probably due to the turbulence created by the vane shaft which was larger than would be used normally and interfered with the smooth flow of the air over the vane.

The effect of placing an obstruction on the back end of the vane is to move the centre of effort further aft. Rough tests indicated that this moved back to about 38 per cent. At this point there is little theoretical advantage in cutting a gap between the vane and the axis.

Our grandfathers, who were not alto-gether stupid, used a vane like that in Fig 32 when they wanted to turn their wind-mills.

This shape gave reasonable strength, enough area and a long lever arm. We would probably use the same shape on our boat's self-steering gears but for one or two snags the foremost of which are

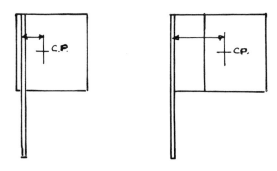

Fig 31 The same vane if placed well away from the vane shaft has a much better torque than if the shaft were actually on the surface of the vane.

Fig 32 Our grandfathers knew the best shape for a wind vane.

Flat Vane: Effect of Removing Part of Vane Next to the Pivot, where Vane Chord = 100m and Span is L

Cut-out	Area mm^2	Lever Arm mm	Lever × Area mm^3
Full Vane	Lx100	33.3	$33.3 \times 100 = 3,330$
10%	Lx 90	40.0	$40 \times 90 = 3,600$
20%	Lx 80	46.7	$46.7 \times 80 = 3,736$
30%	Lx 70	53.3	$53.3 \times 70 = 3,731$
40%	Lx 60	60.0	$60 \times 60 = 3,600$
50%	Lx 50	66.7	$66.7 \times 50 = 3,335$

backstays, which effectively limit the length of lever arms we are able to use (Fig 33).

This consideration, to a certain degree, also controls the shape of the vane which tends to be wider at the top than the bottom. To get the vane as high up as possible is an advantage, as the air is not so subject to obstruction by cabins or helmsman and also the air velocity increases quite considerably a few feet above the water. Strength considerations normally control the height we are able to use.

Model-yacht vanes which steer the rudder directly have to be as efficient as possible and tend to have this shape (Fig 34).

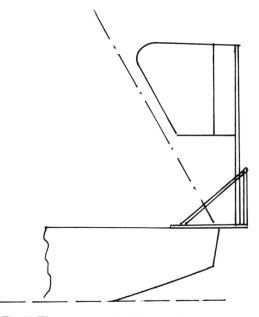

Fig 33 There are practical limitations on a boat which govern a wind-vane shape.

Materials for Construction

Wind vanes are either constructed of cloth stretched over a light frame or from plywood. No other material seems to give a comparable result for either cost or weight. It is essential that the vane be kept as light as possible and practice indicates that, for a plain, flat blade,

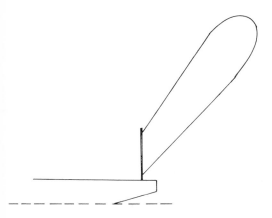

Fig 34 Model yachts use some very pretty wind vanes but they would be weak if increased to full size.

4mm-thick plywood, if suitably stiffened by strips of wood glued on where necessary, is ample.

Balance Weight

Considerable care should always be taken over the design as obviously we cannot expect a heavy blade to be turned by the gentlest of breezes. If the vane is not adequate for the job, aluminium-sheet ears can be added. These will increase the vane's effect considerably.

The balance weight should just balance the weight of the vane about its axis. This is so that when the boat heels over the vane will not be deflected by its own weight and give quite false signals to the rest of the gear (Fig 35). It may be found under test that slight under or over balance gives better results. The reason for this is that, as the boat heels, the hull shape results in the boat turning to weather. Thus if the boat is on the dead run and rolling, the course tends to be serpentine rather than the straight line we wish to see.

As the boat rolls and the vane is slightly out of balance, the vane will turn due to the imbalance weight and this movement will be transmitted to the rudder. Correct imbalance will apply a

corrective to the boat before the usual procedure where the boat goes off course, then the vane senses the wind shift and applies rudder. With an unbalanced vane, correction is applied before the boat starts to swing; with a perfectly balanced vane the correction is not applied until after the boat's heading has already changed.

Unfortunately, the effect of the unbalanced vane is in one direction when running and the other on the beat. This, however, is not a major obstacle as on the beat there is plenty of wind to override the vane's weight. All experiments should be done with the boat on the run when the relative wind is small. If the boat behaves well on the run the much stronger wind on the beat will prevent any wandering on that course.

The lightest vane is probably one made of cloth stretched over a metal frame. The value of cloth on a flat vane is debatable, but for the V-shaped vane the material comes into its own and makes a very neat job (Fig 36). The actual sizes required are discussed in Part Two.

Horizontally Pivoted Vanes

This type of vane, invented by the designer M. Gianoli, was introduced in 1962 and used by Eric Tabarly of France in the 1964 trans-Atlantic race. (This vane, like Columbus' egg, was only waiting for an inventive mind to produce the first.) As soon as the principle was appreciated, there were many others who adopted this type of vane because of its many advantages.

The first advantage is its shape. The shape of a vertical-axis vane is largely governed by the presence of backstays or other obstructions. The horizontal-axis vane can, within strength limits, be as long as you like (Figs 37, 38).

Now to consider the torque produced by the two types of vane. If the vanes are

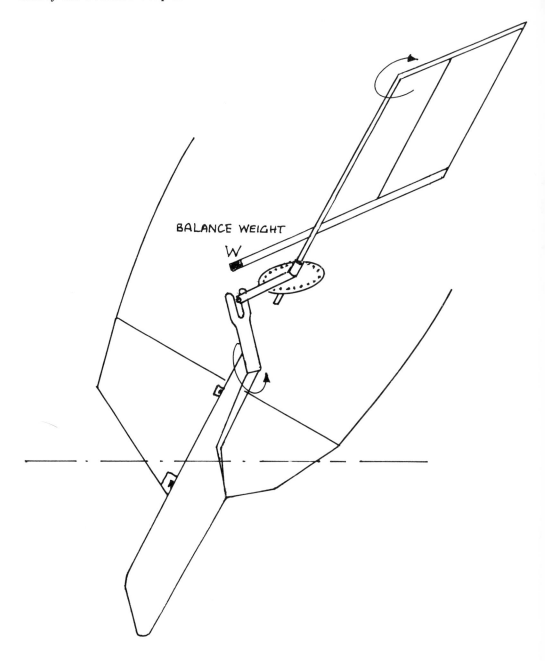

BALANCE WEIGHT

W

Fig 35 Correct balance-weighting (w) of a wind vane improves performance, particularly downwind.

of equal area they will be approximately of the shapes shown. It is clear that the centre of effort of the vertical-axis vane is very much nearer to the vane pivot, in spite of using a gap between vane and axis, than the equivalent horizontal vane. This gives a factor of about 2 in favour of the horizontal-axis vane.

When the wind hits the horizontal-axis vane the vane is forced over, but the angle of it to the wind does not change. In practice, the horizontal-axis vane can be

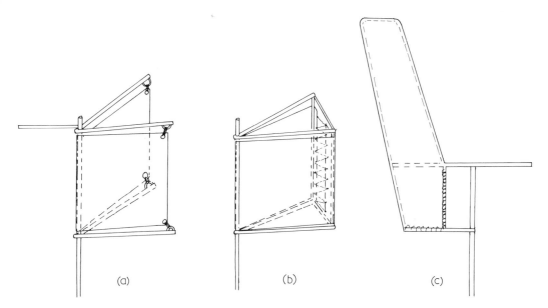

Fig 36 Types of cloth-covered wind vanes. V-shaped as in (a) or (b); a flat vane at (c). The cloth should be stretchless and laced tightly.

geared down in the linkage between the vane and the rudder so that a 45° movement of the vane (about the maximum as after this the vane's power falls off considerably and the mechanical linkage becomes difficult) gives a 15° movement of the rudder, which is usually somewhere near the maximum effective angle to the water flow. The vertical-axis vanes, in practice, have a ratio of vane movement to rudder or trim tab movement of 1:1. This gives the horizontal-axis vane a further advantage of 3:1.

Assume that our gear has to apply a 5° rudder deflection to cope with the boat going off course, then our vertical-axis vane will have to move 5°. The trouble is that in using this 5° the vane moves away from the wind and loses power and the only way this can be corrected is for the boat's course to alter (Fig 39). With the horizontal-axis vane a 5° correction of the rudder requires a 15° movement about the horizontal axis of the vane, but the wind pressure on the vane remains the same, as the angle of the vane to the wind has not altered. We are now trying to

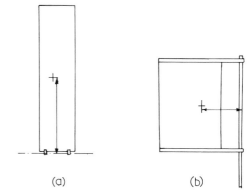

Fig 37 The horizontal-axis vane (a) has a shape advantage over even a well designed vertical-axis vane (b).

compare chalk and cheese, but in practice we seem to get a further benefit of about 2:1 from this effect. Thus, overall, the benefit of a horizontal-axis vane is about 12:1 against the vertical axis vane. Various writers quote anything from 10 to 27.

Before the reader rushes off and produces a postage-stamp sized wind vane, it must be appreciated that the torque of a wind vane is the area multiplied by the distance of the centre of pressure from the axis. This distance, to keep the same shaped vane, is proportional to the size of the vane. Thus, the

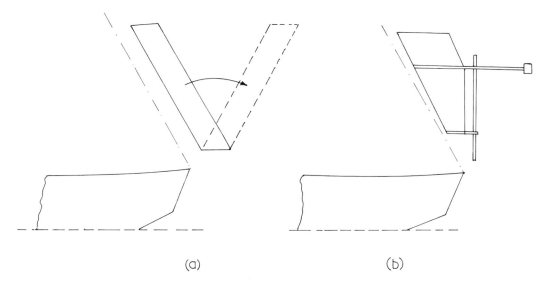

*Fig 38 The horizontal-axis vane (a) can be very
long compared with the vertical-axis vane (b),
which is limited by space considerations.*

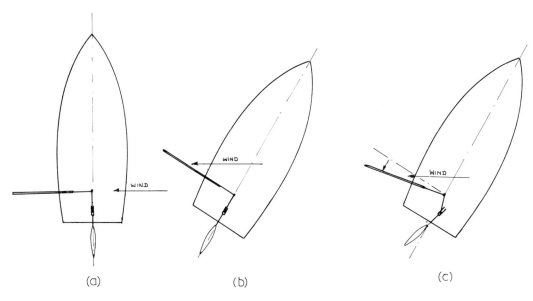

*Fig 39 (a) On a steady course with the vane in
line with the wind; (b) the boat goes off course
putting the vane at an angle to the wind; (c) as
the vane applies rudder, the angle of the vane
to the wind is reduced and so the force on the
vane is diminished.*

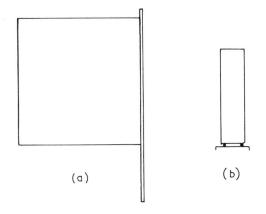

(a)   (b)

Fig 40 An approximate comparison of the respective sizes of (a) a vertically pivoted and (b) a horizontally pivoted vane to give the same result.

power of a vane is proportional to the cube of the vane size (if by size we mean one of the linear dimensions). If the horizontal axis vane has an advantage of 12:1 then the lineal dimensions of the vane can be in the ratio of 1:2.3 comparing it with a vertical axis vane. Fig 40 shows very approximately the size of the two types of vane to produce equal results.

Where a horizontal-axis vane is used for a servo system or to work a trim tab, it can be very much smaller than its vertical-axis equivalent. Where a vane is required to work the boat's tiller direct, a horizontal-axis vane of convenient size may do the job, whereas a vertical-axis vane to produce the same effect would be far too large. My own direct-to-rudder gear uses a vane of 1.5m × 381mm. The mind boggles at using a vertical-axis vane about 3m high and 1m wide.

A further advantage of the horizontal axis is in the counter-balance weight. If a vertical vane is deflected under light wind conditions the rudder will be deflected, and will stay in that position until there is enough force on the vane to return the rudder to central again. With the horizontal axis, the vane can be slightly

over-weighted. A small breeze will deflect the vane, but when this dies the vane will return itself and the rudder to central by the weight of the counterweight itself. This is, in effect, a form of self feed-back. This property is also useful on the run where the correct amount of imbalance of the balance weight can work wonders on the steering.

With all these advantages the reader may well ask why use vertical vanes at all. The answer lies in the greater simplicity of the linkage between vane and rudder, and in the somewhat easier course-setting of the vertical-axis vane. Also the vertical-axis vane, when not in use, can be disconnected from the linkage and will quite happily swing to the wind. With the horizontal vane, the vane will not weathercock into wind and may damage itself by swinging over from one stop to the other, often with considerable force. This means that, when not in use, the vane has to be dismounted and stowed in the boat or lashed in position.

Like vertical-axis vanes, the material for horizontal-axis construction can either be fabric stretched over a frame, or plywood. If plywood, the same thickness, 4mm, can be used if stiffened. If fabric is used then, as the frame is small, quite light-gauge tubing will make a good job and when the gear is not in use it is easy enough to unlace and stow the fabric cover (Fig 41).

Various designers have produced vanes which have their axes not horizontal, but inclined to the horizontal at an angle of between 7° and 25°. In theory, this gives a feed-back effect with the vane losing its angle to the wind as it is deflected. Experiments I made failed to show any real benefit with the vane axis inclined and, taking the extra complication of transmission into consideration, there seems to be no practical reason to use a vane other than one with a truly horizontal axis.

Where the vane has to drive the rudder

direct, the horizontal-axis vane is the only one which will do the job. Where the power of the vane is increased either by a servo system or the use of a trim tab on a balanced auxiliary rudder the vane, if vertical, does not have to be very large and the easier mechanics of the linkage and course setting may tilt the designer's choice towards a vertical-axis vane rather than to the considerably smaller, and therefore possibly more aesthetic-looking, horizontal vane.

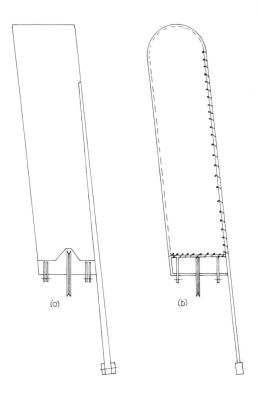

Fig 41 Horizontal-axis vanes either plywood (a) or a cloth-covered frame (b) should be comparatively long and narrow, and sloped to bring the centre of effort of the wind on the vane over the course-setting axis.

CHAPTER FOUR

Course-Setting

Vertically Pivoted Vanes

Course-setting with the vertical vane is merely a matter of rotating the vane relative to the system on which it works and fixing it in the new position.

The simplest form of adjustment is just a clamping screw (Fig 42). This is cheap and easy to make, and course-setting consists of nothing more than loosening the clamping screw, rotating the vane and fastening the screw again. But the system has every disadvantage and should not be used. The fact that a number of the type are sold commercially does not show it is a good method, but does prove that the designer has never leant over the back of his pushpit in a gale trying to loosen a rusted-in screw.

The system is not positive. A gust of wind could easily override the force of the clamping screw and change the setting, besides which there is no indication as to where the vane was before the screw was loosened. If for instance you require a 10° alteration of course, you need to move the vane about 10°; but as soon as you undo the clamping screw you lose your original setting and don't know where to clamp the screw to set it anew. Rapid disengagement is not possible in an emergency.

Better is the disc and clamp. With this system the clamp works on the periphery of a disc which gives it a much better grip. The disc can be marked so that the setting of the vane can be noted, and the course can be set or altered to fine limits (Fig 43).

Still, leaning over the pushpit in a

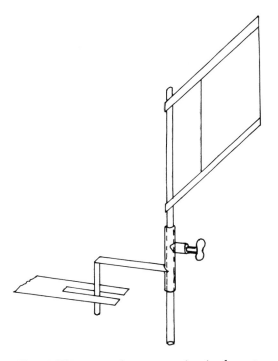

Fig 42 This type of course-setting is cheap to make, but not recommended.

storm to adjust the gear is not really anyone's idea of pleasure and can be dangerous.

A variant of this system which is better still, but which has the disadvantage of working in steps, is to drill the disc with as many holes as is practicable and engage the lever arm by means of a pin. The accuracy of course-setting is limited by the number of holes in the disc which, practically, is only about 48 (Fig 44). If a finer setting is required some type of vernier system can be used, as shown in Fig 45a. The disc has 36 holes on a

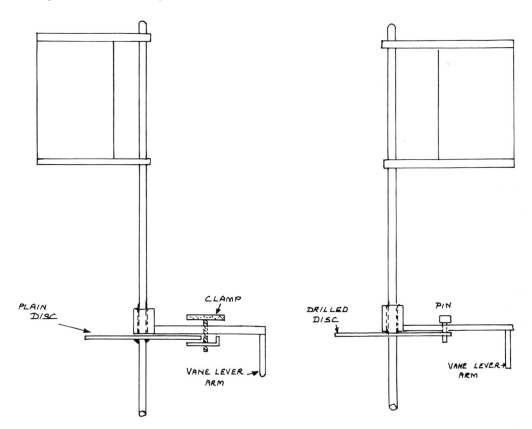

Fig 43 *A better type of course-setting. The clamp is more positive and the disc can be marked to indicate the position of the vane relative to its lever arm.*

Fig 44 *A pin with a drilled disc is an improvement, but the number of holes and thus the course-setting accuracy is limited.*

150mm pitch diameter. The vernier has 5 holes at 12½° pitch. This gives a course setting to 2½°. Fig 45b shows a system of intermediate accuracy. If there are two sets of holes with 48 holes on each diameter, then the course can be set to 3.75°.

In practice, very fine course-setting is not imperative as slight adjustment of the sails will result in more or less rudder being required. This increase or decrease of pressure on the rudder will be provided by the vane being at a different angle to the relative wind and this can be used to adjust the course. In simpler language, the boat can be made to luff up to windward or bear away a bit by the sail adjustment, this in spite of the actual setting of the steering gear.

An improvement of the pin and disc is the latch (Fig 46). Diagram 46a consists of our pin and disc, but with the pin fixed to the lever arm and the lever arm free to be raised by pulling on a line. This type is disengaged by lifting the lever arm until the pin is clear of the disc. Type 46b does not use a disc, but a toothed wheel with a latch which drops into one of the gear teeth. These two systems are both alike. The choice of a drilled disc or toothed wheel depends upon which is easier or cheaper to produce.

Both of these systems have instant disengagement by means of the control line and, as there is no limit to the length of a piece of string, the control can be placed anywhere handy in the cockpit. No longer is it dangerous to change course. To set

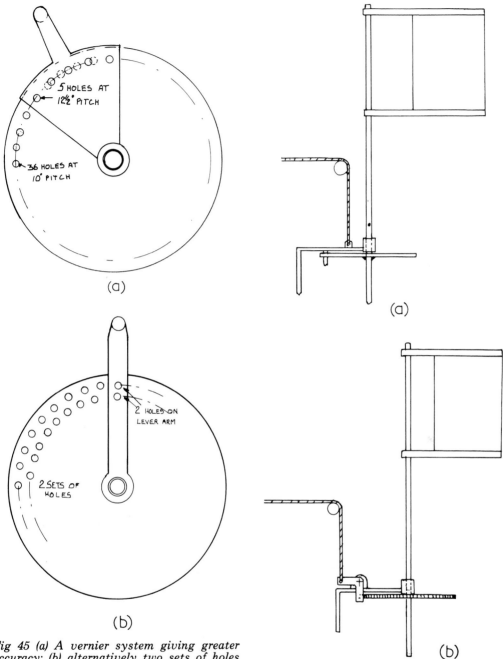

(a)

5 HOLES AT 12½° PITCH

36 HOLES AT 10° PITCH

2 HOLES ON LEVER ARM

2 SETS OF HOLES

(b)

(a)

(b)

Fig 45 (a) A vernier system giving greater accuracy; (b) alternatively two sets of holes can be used.

Fig 46 (a) A simple latch system using a drilled disc; (b) a toothed wheel in place of the disc.

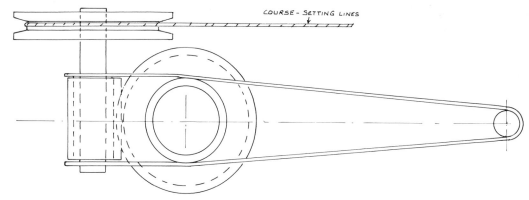

COURSE - SETTING LINES

WORM   12 DIAMETRAL PITCH
25mm (1") P.C.D SINGLE START
WHEEL 12 DIAMETRAL PITCH
62.5mm (2½") P.C.D.

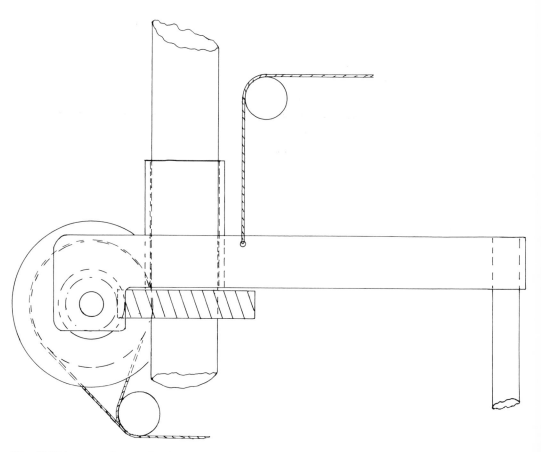

Fig 47 *This type of worm latch similar to that used on the Hasler gears gives infinitely variable adjustment and immediate disengagement.*

the course, the vane is disengaged; the boat is then sailed, using the boat's tiller or wheel, onto the course required or, in practice, slightly below the true course and then the latch is dropped in. The self-steering will then take over with the boat luffing up until there is enough wind pressure on the vane to provide the force to control the rudder. Minor adjustments of course can be made by sail adjustment.

There is one snag with this system, namely that if we are using our sails to effect minor course variations we are not sailing to our best advantage.

The course-setting method adopted by Hasler for his vertical-vane gears has all we want. The latch of the simple disc and latch system is replaced by a worm wheel (Fig 47). With this system the gear can be disengaged by pulling the control line which raises the worm wheel out of gear. When the line is released the worm falls into gear with the wheel fixing the lever relative to the vane. Minor course alterations are made by turning the worm by means of course-setting lines round a pulley, these course-setting lines being led downwards so that the pull on the lines counteracts the tendency of the worm to lift up and disengage itself as the worm is rotated.

Where large alterations of course are required, such as a tack, the worm latch is disengaged, the boat steered by hand onto the new course and the latch dropped back. Minor alterations are then done by the worm.

Horizontally Pivoted Vanes

In the case of the horizontal-axis vane, course-setting is obtained by rotating the whole vane assembly about the vertical axis (Fig 48).

At its simplest the vane is turned by hand. If remote control is required then a groove cut round the vane carrier can be used with an endless line to turn the vane from a distance. The carrier also gives a

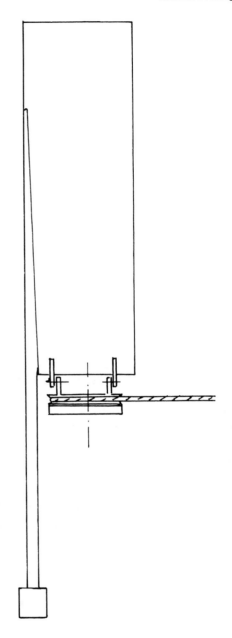

Fig 48 A horizontal-axis vane is rotated about a vertical axis for course-setting.

big pulley to make turning the assembly easy, and cleating the line will hold the vane in the desired position. More sophisticated systems use either a ratchet or worm gear to rotate the vane using course-setting lines from the cockpit.

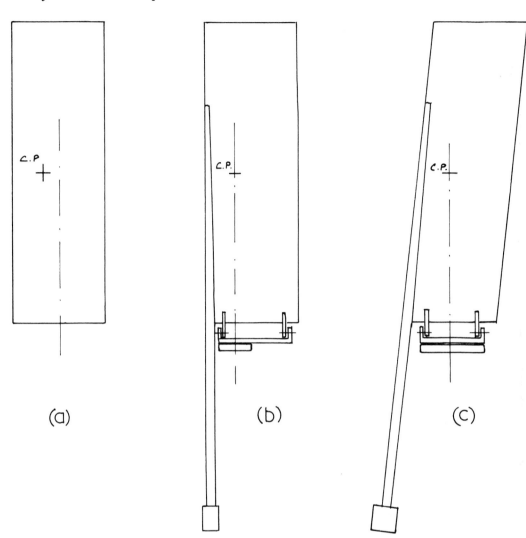

Fig 49 (a) If the centre of pressure of the wind is away from the course-setting axis the vane is hard to turn and tends to turn itself; (b) By having the course-setting axis on the same line as the centre of pressure the vane is easy to turn and, once turned, will stay in position; (c) By sloping the vane the same effect is obtained and it looks better.

The force required to turn the vane for course-setting is proportional to the distance from the centre of pressure of the vane to the course-setting axis. The likelihood of the vane-setting slipping due to a gust of wind is also proportional to this distance (Fig 49a). By offsetting the vane (49b) the two axes lie on the same line and the vane is easy to turn and, once turned, only a small amount of friction will hold it in position. By inclining the vane (49c) a more elegant shape is produced.

CHAPTER FIVE

Linkage and Feed-back

General Description

In its simplest form the linkage is just a means of connecting the wind vane to the rudder or trim tab it is working on. A well designed linkage has however a more important function, that is to give smooth and well controlled steering. Consider how a good helmsman steers a boat. He does not apply full rudder and then wait until the boat is on course before he checks the boat's swing. He rather applies the necessary rudder, but as soon as the boat starts to swing he reduces the rudder until there is just enough to hold the boat on course when the swing is finished.

With properly designed linkage we can introduce 'feed-back', which has something of the same effect and helps the gear steer a good course without violent oversteer with the boat swinging from side to side of the required course and never seeming to settle down. It is true that some boats are well designed and prefer to go in a straight line, but others have no natural directional stability and never seem to settle down. I once travelled on a 57ft boat on which we spent about a week trying to get her to steer herself by adjusting the sails. Finally, we succeeded and the boat sailed 'hands off' for about ten seconds. *Josephine* on the other hand would hold a course if you just let the tiller go. If you lashed the tiller she hardly needed self-steering, but then *Josephine* was designed to sail and not to fit into a rating rule.

The effect of feed-back can be described

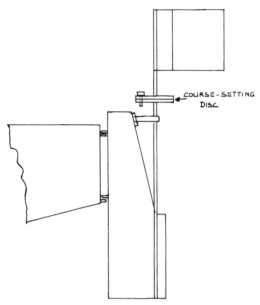

Fig 50 A trim tab with the vane directly coupled is cheap and simple to make but is not recommended unless the boat is directionally stable.

by practical illustration using a simple vertical-axis vane and a rudder with a trim tab. The vane of the trim tab set-up rotates in the same direction as the trim tab, so in its simplest form we can make a gear (Fig 50).

This is very simple, cheap to make and, you would think, should be ideal, until its action is considered step by step to see what happens (Fig 51). For convenience we are now looking down on the gears, and in all these illustrations the vane is showing pointing direct into wind. This does not happen exactly (as we have already explained) but if we make this assumption it simplifies the matter.

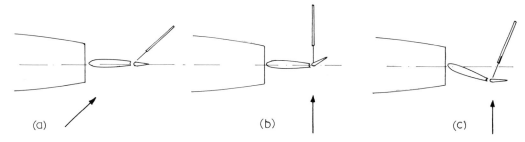

Fig 51 (a) A trim-tab gear as shown in Fig 50 cruising along; (b) A change of wind deflects the trim tab; (c) As the rudder swings over it increases the angle of the vane to the wind which then applies more trim tab.

It will be seen that as the rudder moves it automatically increases the amount of trim tab and this again increases the rudder deflection. Contrary to some published opinion, this effect, which is caused by the angular rotation of the rudder producing the same angular rotation of the trim tab, will apply regardless of the position of the rudder and trim-tab axes.

This whole set-up is unstable and, unless the boat herself is very stable and can override the instability of the vane gear, steering is going to be very erratic. In this case, where the effect of the rudder movement is to increase the effect of the wind vane, we say the gear has positive feed-back, that is to say, the effect of the rudder movement is to increase the tab forces acting on it.

Trim tab with Linkage

Now let us consider the same arrangement of rudder and trim tab, but with a linkage as shown in Fig 52.

We have not shown how the vane is supported, but this makes the picture easier to appreciate (and draw).

Fig 53 is looking down on the gear with the boat cruising along (a). A wind shift (b) deflects the vane and trim tab. This causes the rudder to swing over until the trim tab is nearly in line with the rudder

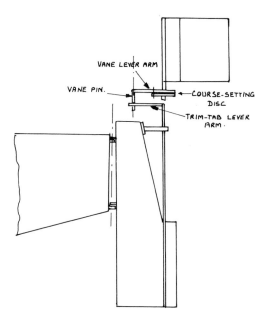

VANE LEVER ARM

VANE PIN.

COURSE-SETTING DISC

TRIM-TAB LEVER ARM.

Fig 52 With the vane connected to the trim tab as shown, more stable steering is obtained.

again, as at (c). Now we have a linkage which behaves like our helmsman. It applies rudder due to a wind shift, but as soon as the rudder starts to work the effect of the vane is reduced. This is described as negative feed-back. Here the rudder movement decreases the trim-tab deflection and consequently the force it applies.

It will be seen (Fig 52) that the trim-tab pin is placed between the trim-tab axis and the rudder axis. If the trim-tab pin is directly on the rudder-pintle axis the movement of the rudder will not affect the trim-tab angle and so there will be no feed-back. If we place the trim-tab pin forward of the rudder pintle then we have

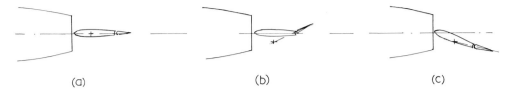

Fig 53 (a) Cruising along with no deflection of trim tab; (b) The vane pin deflects the trim tab; (c) As the rudder swings over the trim tab is straightened relative to the rudder and the power is reduced.

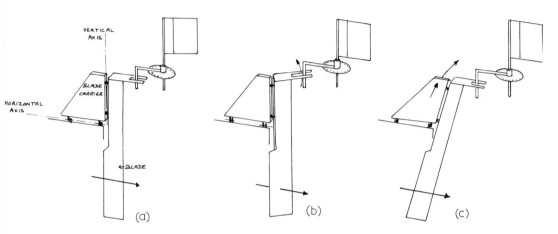

Fig 54 Pendulum servo feed-back:
(a) Cruising; (b) The vane deflects the blade which swings like a pendulum; (c) When the blade swings over, its angle to the water is reduced as is the pull in the tiller lines.

again produced a gear where the vane effect is increased by the rudder movement.

Pendulum servo

With pendulum servo gears the vane works the oar in the reverse direction and so our linkage is of the form shown in Fig 54. In (a) the pendulum servo gear with the servo blade lined up with the water flow and no pull on the tiller lines. In (b) the effect of a wind shift (or the boat going off course). The wind vane is deflected and this moves the blade about the blade carrier's vertical axis. The blade inclined at an angle to the passing water flow creates a very powerful force which

will swing the blade, acting like a pendulum, about the blade carrier's horizontal axis. This moves the tiller lines and, thus, the boat's tiller or wheel. As the blade swings over like a pendulum the effect, if the wind vane stays lined up with the new wind, is to gradually straighten out the blade and decrease the force which it exerts on the tiller lines (c). This is a stable situation. When the blade swings over to the end of its travel, its deflection is reduced to zero and all sideways force is taken off it, which prevents breakage as the blade does not hit the end of its travel under full power. If the vane pin acted not above the blade carrier's axis, but exactly on the axis, then the swing of the blade would not alter the angle of the blade relative to the water flow. In this condition there would be no feed-back. If the vane pin acted below the axis, we would have positive feed-back and instability of the whole

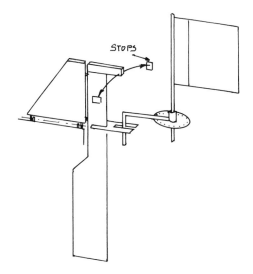

Fig 56 Stability of steering can be obtained by the use of stops which straighten out the blade as it swings over on its horizontal axis. This has the further function of aligning the blade with the water flow at the end of the servo-blade sweep thus eliminating all strain on the blade.

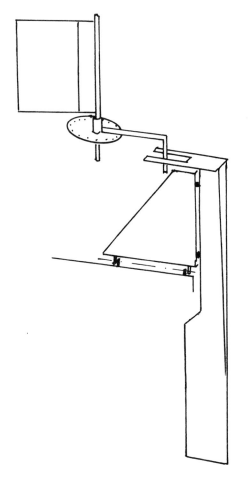

Fig 55 This feed-back effect is only obtainable when the lever is in the correct position in relation to the blade and carrier bearings. As shown the feed-back is positive and over-steer almost certain.

system. The feed-back effect would be reversed if the vane was on the other side of the blade (Fig 55). In these illustrations the feed-back is probably excessive, but is shown thus for clarity.

Two-stage System and Feed-back

There are two interesting points which emerge from our feed-back examples. The first is that, in order to obtain feed-back, we must have a two-stage system. With the trim-tab set-up, the vane deflects the trim tab, the trim tab deflects the rudder and the rudder deflection acts back to

reduce the effectiveness of the tab movement. With the pendulum servo, the vane deflects the blade, the blade swings on the blade carrier's horizontal axis and this deflection is used to reduce the effectiveness of the vane movement.

Where a vane works direct onto a rudder no geometrical feed-back is possible. If there is no two-stage operation we must abandon geometrical linkages for weights, shock cord, or stops which provide a form of synthetic feed-back. The action of a vertical-axis vane can, perhaps, be considered an exception to this rule as, when the vane is deflected, the force of the wind on the vane is reduced. Thus it has its own built in feed-back.

Fig 56 shows a pendulum servo gear with the vane lever acting on the blade on the horizontal axis of the blade carrier. The effect of the blade carrier swinging from side to side will have no effect on the deflection the vane gives the blade. There is no geometrical feed-back. As, however, the blade swings over to one side, the

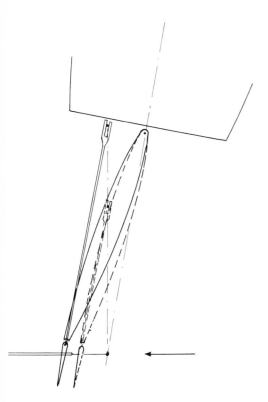

Fig 57 Feed-back gives less rudder movement (dotted lines) than if there is no feed-back (full lines), and thus ultimate power is lost to get smoother steering.

blade lever strikes one of the stops and the blade is deflected back to the fore and aft position. There is feed-back from stops.

Feed-back and Course Accuracy

The other important point is that feed-back, whilst giving smoother steering, results in the boat wandering off course rather more as the wind speed changes and there is a demand for more or less rudder to counteract the resultant weather helm. This is best illustrated by an example of a trim-tab rudder (Fig 57a). The rudder with no feed-back, if its trim tab is deflected by the vane, will swing over, forced by the full power of the tab. If, however, the linkage is designed for feed-back (dotted lines) then, as the

rudder swings over, the effectiveness of the vane on the trim tab is reduced and, therefore, the power of the rudder (b). The only way we can get enough power on the rudder is to sail off course, when the extra vane deflection will override the effect of the feed-back. With feed-back we will get a more stable ride, but at the expense of greater course variations with change of wind strength.

Synthetic feed-back

There are several examples of synthetic feed-back which we can use either in conjunction with true geometric feed-back or on their own. The earliest to be used was the weighted rudder of model boats. Here, as the boat leant over due to the wind pressure, the boat tended to develop weather helm and luff up into the wind. But as the boat heeled, the rudder was no longer vertical and the weight on the rudder applied automatic weather helm (Fig 58).

The vane, if on a horizontal axis, can be over-weighted so that, as the vane swings over, the effect of the wind is reduced as the vane is inclined away from the wind, but the effect of the excess weight is increased as the vane is tilted (Fig 59).

The weight does not affect small vane movements, but when the swing is longer the weight serves to reduce the vane's effectiveness. Similarly, if a vertical-axis vane is used we can use shock cord to restrict the vane's effectiveness by placing it so that it has the same effect on the vertical vane as the weight on a horizontal vane (Fig 60).

Examples of Linkage

The actual method of joining a vertical vane to a trim tab is based on our need to have the vane and trim tab rotating the same way; but if the vane works a rudder direct, or a pendulum servo blade, the vane and the rudder or blade rotate in

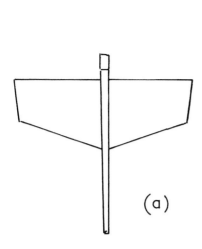 (a)

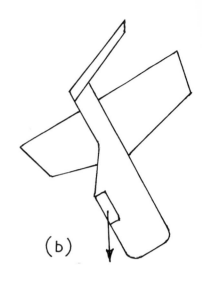 (b)

Fig 58 Synthetic feed-back can be obtained by weights on the rudder, as was done on early model yachts: (a) with the boat upright no rudder (b) as the boat heels rudder is applied by the weight.

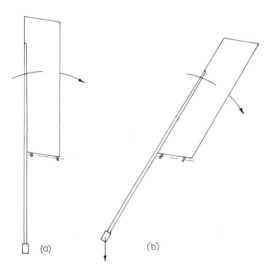

(a)   (b)

Fig 59 Over-weighting a horizontal-axis vane (a) gives the effect of feed-back. As the vane 'flops' over (b), the area exposed to the wind is reduced, but the effect of the overbalance weight is increased. If the wind dies, the vane will return to the vertical position and straighten the rudder.

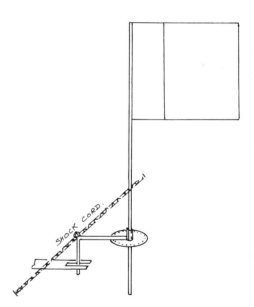

Fig 60 Applying shock cord each side of the vane lever arm has the same effect as over-weighting a horizontal-axis vane.

opposite directions (Fig 61). In the early stages of self-steering development, when the trim tab was the only means of power amplification, all sorts of weird and wonderful systems of transmission between vane and trim tab were adopted. Universal joints, underwater lever arms and hollow rudder stocks were all used — entailing underwater mechanisms which were subject to corrosion, weed growth and jamming by marine organisms, and the boat being slipped before repairs

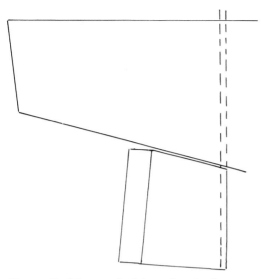

Fig 61 Rudders tucked in under the boat are not suitable for trim tabs.

could be effective. Today, if the boat's rudder is tucked in under the boat then either a pendulum servo, or an independent auxiliary rudder, or a Saye's system is fitted. No attempt should be made to fit a trim tab where the linkage between vane and tab is at all complicated.

Fig 62 gives the simplest arrangement of a trim tab fitted to an existing rudder with the vane acting at about 80 per cent of the distance between the trim tab and rudder axes. The disadvantage here is that the vane has to be mounted quite a bit outboard of the boat. Provided, however, we still keep the trim-tab lever arm of the same effective length, there are alternative methods of mounting the vane.

Fig 63 shows alternative systems which have the advantage of bringing the vane inboard where it can be attached to the pushpit, rather than having to construct a vane support outboard of the boat's stern. The arrangements at (a) and (b) are not quite perfect as the feed-back ratio is not constant, but the difference is not great and can be accepted because of

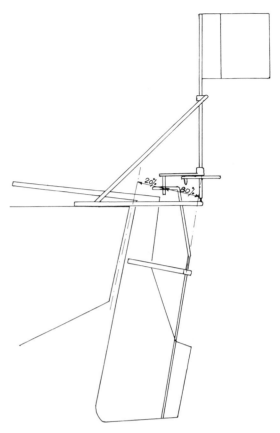

Fig 62 Straightforward arrangement of trim-tab fitted to an outboard rudder.

the easier vane mounting.

A plain fork and pin connection (Fig 64) is simple and gives no trouble, although it may be noisy. Placing a plastic sleeve over the pin will usually cure this, although spares should be carried as wear may be rapid. The connection may not be suitable if the axis of the trim tab is at too great an angle to the axis of the vane. With too great an angle between the fork and the pin, the fork has to be widened to take the pin and this may result in too much slop in the linkage. Pendulum servo gears, where the servo-blade sweep is up to 22½ degrees, give no trouble with a fork and pin linkage. If however a trim-tab gear is required with the tab axis at, say, 45° to the vane axis, a more sophisticated linkage using universal

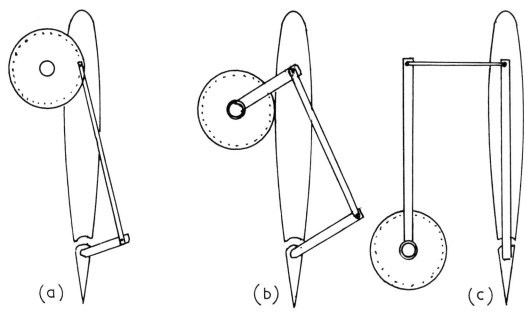

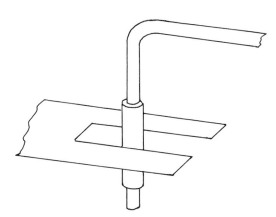

Fig 63 Methods (a), (b), (c) of working the trim tab so that the vane is not too far outboard of the boat. Note in all cases there will be feedback but the amount will depend on the actual geometry chosen.

joints is probably required. Where an intermediate link is required this can either be with ball and socket ends, or the following far simpler system which gives long life at low cost.

Fig 65a shows a simple, low cost link made of 8mm diameter steel. It is strong enough for the job and, although not so sophisticated as the ball and socket joints, will not rust up in sea water. It is also easily replaceable. One end of the connection rod should be drilled for a split pin so that your crew cannot throw it overboard; the other end can be plain if used with a drilled disc, or can also have a split pin if used to connect two lever arms. If necessary the link can have a bottle screw in the middle to make it adjustable (Fig 65b). This type of link can be used with a latch system (Fig 66).

With all vane linkages care should be taken that the linkage cannot travel over dead centre. Fig 63 showed simple links between a vane and trim tab. If, due to a violent gust of wind or sudden wind shift

Fig 64 A simple fork and pin connection.

the vane rotates too far, the linkage will travel over the dead centre of the vane disc and the effect of the vane rotation on the trim tab will be reversed. This can be prevented by suitable stops on either the vane shaft or the link to limit the link travel.

The question of the ratio of vane movement to tab or rudder movement has been investigated by all researchers with no very definite results emerging. For practical purposes a ratio of vertical vane to trim tab, or balanced rudder, of 1:1 gives good average results although

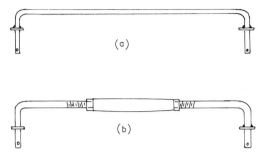

Fig 65 Simple links: (a) plain; (b) with bottle screw to make it adjustable.

builders of their own gears may like to experiment with their own variations. Where the vane is on a horizontal axis and works a balanced rudder, a ratio of 3:1 gives the best results, although this ratio is partly dependent on the respective areas of vane and rudder. As a guide, 0.36sq metres of vane (1,200mm × 300mm) linking a rudder with 300mm chord (0.27sq metres immersed in the water), with pintles at 23 per cent aft of the leading edge, gave the best results with a 3:1 ratio. With this ratio the wind vane is balanced by the water pressure on the rudder. The balance is obviously dependent on the wind and boat speed, and can only be approximate as the ratio of wind to boat speed is a very variable factor.

In some steering gears, motion from the vane to the rudder is transmitted by bowden cable. This method, although more subject to friction or breaking of the cable can nevertheless be used when the vane is to be placed at a distance from the rudder. For this purpose the cable can be long, or go round corners.

The linkage for a horizontal-axis vane to a rudder or trim tab is slightly more complicated, as the motion has to be transmitted from the horizontal axis of the vane to the vertical axis of the rudder. Just using cords is quite efficient and was the method adopted by the Amateur Yacht Research Society (AYRS) for their simple gear. The drive shown in Fig 67 is

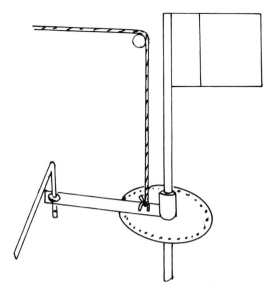

Fig 66 Latch mechanism for use with simple links.

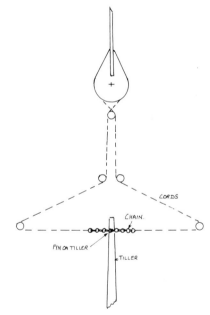

Fig 67 The simplest method of linkage between a horizontal-axis vane and the boat's tiller, using lines and a chain and pin connection.

a development of their early ideas; it gives good results if the cords are of small diameter and with the minimum of stretch. The pulleys should be large to avoid friction and the ends of the cords

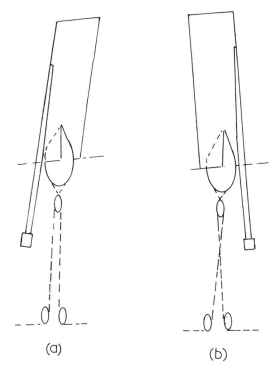

(a)   (b)

Fig 68 This line linkage (a) can only be rotated 180° each way for course-setting before the lines get crossed (b).

should be tied to a piece of chain. The system is engaged with the tiller or trim-tab lever by the chain being dropped over a pin. The same system, using thin wire in place of cord, was used by Gunning in his successful production gear. In his case the vane worked a trim tab attached to a skeg of a pendulum servo system.

All these line linkages, although simple and cheap to construct, have one fundamental drawback: they can be rotated only 180° before the lines get tangled up with each other (Fig 68). If the gear is set for beating on one tack, a new tack is not just a matter of turning the vane by a comparatively small amount, but the vane has to be rotated the long way round.

A better method of transmission was developed by Gianoli for use on MNOP gears and first brought to prominence when used by Tabarly for the 1964

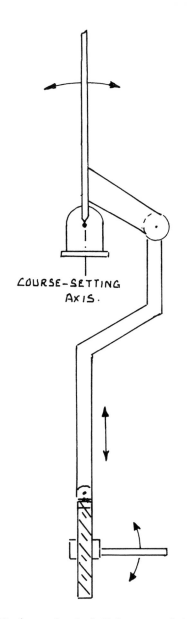

Fig 69 A mechanical linkage used by M. Gianoli; there is no limitation on the vane rotation for course-setting.

Observer Single-handed Trans-Atlantic race (OSTA), unfortunately before the bugs had been ironed out of it. In this system the vane, instead of being fitted with a line pulley, had a lever arm to which was attached a curved connecting rod which went down the centre of the vane support and had a swivel at its lower

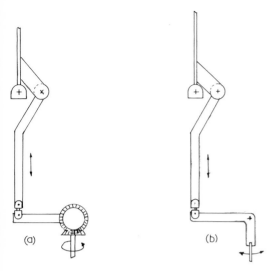

(a)   (b)

Fig 70 The same mechanical linkage can work either a gear (a) or a lever arm (b) to get the rotation wanted for the rudder.

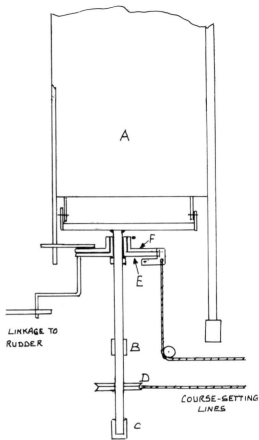

LINKAGE TO RUDDER

COURSE-SETTING LINES

Fig 71 An alternative system of vane to rudder linkage; fairly neat but a little complicated.

end. Thus the vane could be rotated in any direction with no up and down motion of the connecting rod — provided, of course, the swivel was dead on the axis of rotation of the course-setting. As soon as the vane was moved by the wind, the vane motion was transposed to an up and down movement of the connecting rod. Gianoli used this up and down motion to work a 'Yankee screwdriver' mechanism which gave him the rotation he required for his trim tab (Fig 69).

This same idea has been used by a number of designers, some using a gear train and others a series of levers to get the final output they want (Fig 70). Both these systems, if well designed, have worked well; and the actual method chosen would seem to depend on the designer's personal ideas, or the capacity of the workshop which turns them out.

A system which has not yet been developed commercially, but which is simple and neat and can be used with remote control, is shown in Fig 71.

A horizontal-axis vane A is fixed to a shaft which is free to rotate in bearings B and C. A grooved pulley D with control lines can turn the vane assembly and fix it in position relative to the wind. An arm E which rotates freely on the shaft has at one end a lever arm to work a rudder or trim tab, and at the other a latch. A toothed wheel F is free to rotate on the spigot on E, and has a fork engaging with a pin on the wind vane. The latch engages with the toothed wheel. To set course the latch is disengaged by the latch line. The vane is then rotated by the course-setting lines until the vane is pointing into wind. The latch is then engaged. If the latch were fitted with a worm the assembly would then become infinitely variable for wind direction.

CHAPTER SIX

Rudders and Blades

Looking at the curves of Fig 12 (p 23) which show how different shaped foils perform, it will be seen that a high-aspect-ratio rudder gives good lift, but develops this and stalls at 15° to 18° deflection. A low-aspect-ratio rudder does not have a sudden stall, but does not develop full power until the tiller is at an angle of 30° or more. In designing self-steering rudders we obviously want high lift and we do not want to have to turn the rudder too far before it develops its maximum efficiency. Generally, therefore, we use the profile shape which has the greatest aspect ratio possible with other factors, such as strength, taken into consideration. There is also a further factor in favour of high-aspect-ratio foils. They are easier to turn, as the distance from the centre of effort to the pintles is less. It is true that all self-steering rudders should be designed to be balanced, but the high-aspect-ratio foil is always easier to turn (Fig 72).

When we come to pendulum-servo gears a long blade is imperative, otherwise when the boat is hard pressed and the pendulum well over sideways to try to resist the resultant weather helm, the blade may well come nearly out of the water and lose all power. This is clearly shown in Fig 17. As the blade has a lot of leverage on the tiller it is not difficult to design a section which is strong enough for the work it has to do — anyone knows that, when rowing, it is hard to break a well designed oar even with a very muscular oarsman. The part of the servo

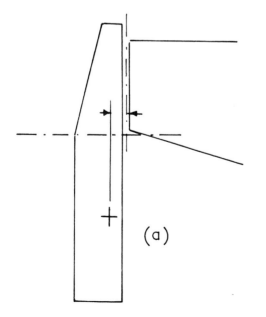

(a)

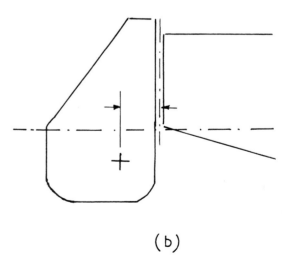

(b)

Fig 72 A high-aspect-ratio rudder (a) is always easier to turn than a low-aspect-ratio one (b).

blade in the water usually has a chord of 100mm to 150mm with an immersed depth of between 450mm and 600mm. This gives an aspect ratio of about 5.

With an auxiliary rudder, either plain balanced or controlled by a trim tab, the question of strength is important as this rudder has to be as strong, proportionately to its area, as the boat's main rudder. For this reason, such rudders have aspect ratios normally between 1½ and 2½. This gives a good rudder which is not too difficult to turn, particularly if well balanced. It is strong enough as it does not have too much unsupported length, and at the same time does not

require too much tiller movement to become effective.

The design of a plain rudder is quite easy as the lift and drag coefficients are well documented and the section using the NACA 00 series has a well defined and stable centre of pressure. With trim tabs, however, the problem is not quite so easy as, although there is plenty of literature on large trim tabs, or flaps used as high-lift devices on aircraft (Fig 73a), there is little covering the field where the foil is at a negative angle of attack (Fig 73b).

Trim Tabs

The earliest tabs were small streamlined foils located aft of the aircraft control surfaces (Fig 74a). Turning them was easy as the foils were small and were also pivoted near their centre of effort. Turning the foil produced a very consider-

Fig 73 The flap on an aircraft wing results in a curved airfoil with good streamline charac-teristics (a). But a trim tab used on a free rudder deflects the rudder so that the tab is working against the rudder and reducing its efficiency (b).

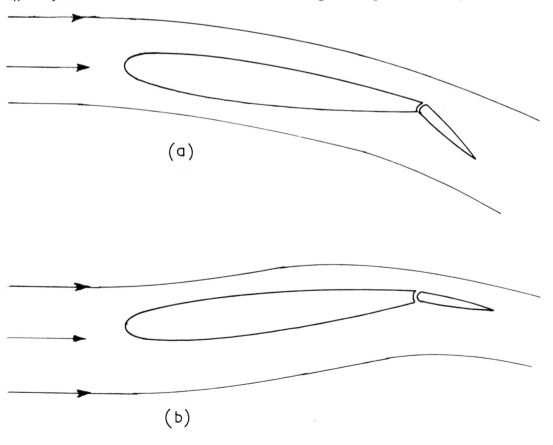

(a)

(b)

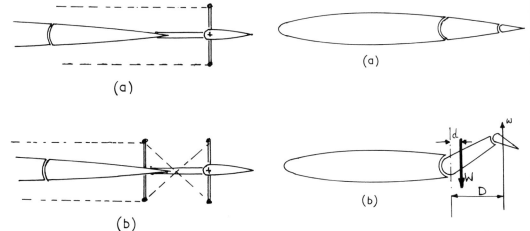

Fig 74 (a) An early trim tab used to work air-craft controls. Here the pilot's controls deflect the trim tab which in turn deflects the aileron; (b) An alternative version where the control wires deflect the control surface but the crossed wires deflect the trim tab in the reversed direction, thus taking the load off the pilot.

Fig 75 (a) The modern control tab. Ailerons and tab are all streamlined into one wing; (b) a small force 'w' on the tab produces a large force on the aileron W.

able force as the slipstream hit the inclined surface; this force then turned the main control surfaces. A variation of this was to have a system as in Fig 74b where the movement of the main surfaces produced a corresponding movement in the foil, which then very much reduced the force the pilot had to use.

This was all a little untidy with bits of string all over the place giving lots of air-drag and the final development was to make the foil, which now became a trim tab, part of the streamline section of the main control (Fig 75a). When the trim tab was turned this produced a thrust which moved the main control surface until there was a balance between the leverage of the main surface and the leverage of the trim tab (Fig 75b).

Although the trim tab is small the distance away from the control pivot is large, whereas the control surface's much larger force has only a small lever arm. The actual effect on the aeroplane's wing is, roughly, the difference between the two forces.

Self-steering rudders of this configuration have been made with the aircraft wing becoming a fixed skeg with a rudder worked by a trim tab (Fig 76). This system has the advantage of strength and lower stresses in the rudder pintles. The rudder is not hard to turn as a considerable amount of lift is provided by the skeg. The rudder will, however, be harder to turn than a fully balanced spade-type rudder and the extra com-plication below the water makes the system a little less reliable.

The system usually adopted today is as shown in Fig 77. Here the auxiliary rudder is mounted so that the axis of the pintles passes very close to the centre of effort; thus the rudder is very easily turned by the trim tab.

The literature on the subject of trim tabs as used for self-steering is a little misleading. A balanced tab working a balanced rudder when the two are separated is amenable to simple calcula-tion as both can be taken as simple foils in clear water. The forces generated by the foils and their centres of effort are well documented and can be used with a fair degree of accuracy. If both foils are of the same shape their centres of pressure will

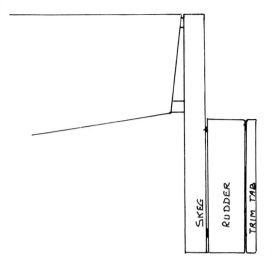

Fig 76 Skeg, rudder and trim tab form a strong independent unit; but it is a little over-complicated and the rudder is not balanced.

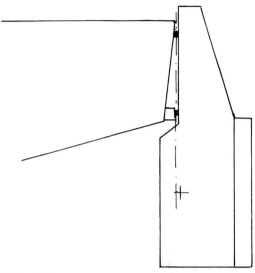

Fig 77 The usual type of independent unit consists of a balanced rudder and trim tab.

be in the same relative position, and the force generated will be proportional to their area. When the trim tab is joined to the foil we have an altogether different set of conditions. The curves in Fig 78 show what happens when a 30 per cent flap is applied at 30°. It will be seen that the pressure per unit area is far larger at the front end of the foil than the back, so the loading per unit area on the tab is less than the loading in the main foil; also the effect of the flap is to increase the effect of the main foil. Results reported in NACA technical report No 938 show that the optimum increment of lift is with a flap of 20 per cent to 25 per cent of chord deflected about 60°. These results were, however, for wings at a positive angle of attack and not like our rudder and trim tab at a negative angle of attack. It seemed that if we were to get any design data which could have any use on a boat it would be necessary to construct an experimental rig. As no test tank was available I decided to use my yacht *Tabitha* as a floating laboratory.

It is fully appreciated that my findings could be in error for a number of reasons which, before someone shoots me down for sloppy thinking, I will enumerate. The

results could only be reproduced if the sea was completely calm and this was not always the case. I could not do the tests alone in enclosed water as my attention would be focussed on the test rig and I would not be looking where I was going; I had to carry out all tests in reasonably open water which was not always dead calm. The tests were carried out under motor and the boat had to be kept as nearly as possible upright to ensure equal immersion of the rudders. This was not always certain as the weight of myself, or fuel, or stores did not always leave the boat perfectly upright. The test rudders were as far away from the propeller slip-stream as possible, but there must have been some effect on the water flow from the propeller rotation. The rudders were in the disturbed water behind the boat; this is a problem with any ship's rudder and self-steering gears are no exception.

With all these disadvantages one might ask why take the trouble to do the tests at all. The answer is that in the kingdom of the blind the partially sighted man is king, and although controlled laboratory experiments would give far more accurate results the findings are approximately

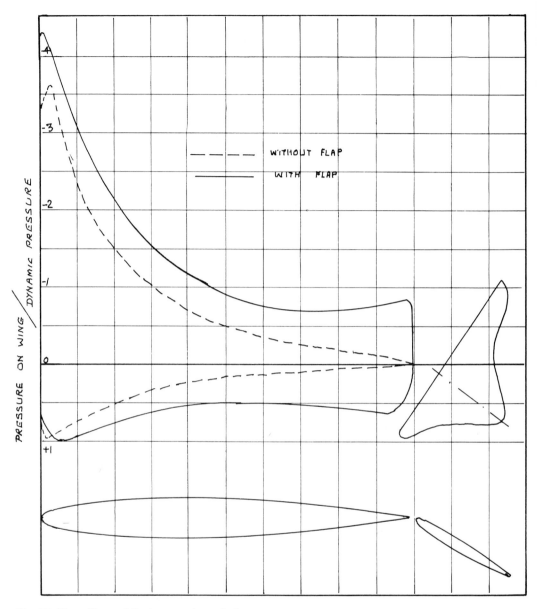

Fig 78 The effect of fitting a trim tab is to increase the lifting power of the main foil, but at negative angles of attack conditions alter.

those which can be expected under normal conditions of sailing.

The test rig was designed to compare a standard rudder against the various configurations about which information was sought. For this purpose the two rudders were mounted as in Fig 79, each being mounted on a frame so that the rudders could be turned in their normal axes. As the frame was also hinged on a horizontal axis, any sideways force generated by the rudder would tend to swing the rudder on this axis. By connecting together the two frames and setting the standard rudder and the test rudder to oppose each other it was easy to determine which was the more powerful, or at which angle of rudder deflection the rudders were exactly balanced.

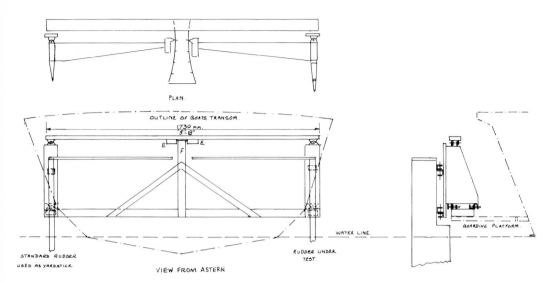

PLAN.

OUTLINE OF BOATS TRANSOM.
1730 mm
5'-8"

E   E
F

WATER LINE.

STANDARD RUDDER
USED AS YARDSTICK.

VIEW FROM ASTERN

RUDDER UNDER
TEST.

BOARDING PLATFORM.

Fig 79 Test rig mounted on Tabitha.

The standard rudder (Fig 80) against which the other configurations were measured was of NACA 0017 section with a chord of 241mm. The test rudder measured 253mm on the chord; to this could be added trim tabs of 43mm, 61mm and 100mm chord, giving tab/chord ratios of 14.5, 19.5 and 28.3 per cent. The balance of the rudder could be changed by packers between the pintles and the rudder body to give 6, 12, 18, 24 and 31 per cent balance. The standard rudder had 22 per cent of balance to make it easy to turn. The immersed depth was approximately 600mm with the boat in motion. This gave an aspect ratio of 2.4:1.

The two rudders were set to balance one another with the arm F oscillating in the stops E. The boat's tiller was then fixed so that the boat was going in a straight line. The two test rudders were then re-set to balance, the turning of the boat upset this balance, and the boat's rudder re-set. By going from one to the other, a balance could be obtained where the test rudders were just balanced and the boat was going straight. As speed made a difference to the results it was very necessary to make frequent checks of the

log, and as this was a Sumlog the speed was probably the least controllable factor of all.

The first test on the rig was to determine how the trim tab worked. For this purpose, the trim tab was pegged at various angles to the rudder and the deflection this caused to the rudder recorded (Fig 81).

Fig 82 gives the result of these tests which were carried out with a 19.5 per cent trim tab, ie a tab whose chord was 19.5 per cent of the cnord of combined rudder and tab. It will be seen that when the rudder is almost perfectly balanced, ie at 24 per cent, a degree of deflection of the trim tab will result in nearly 10° of deflection of the rudder. This is too sensitive for practical purposes and the rudder may shake itself to pieces, as happened with an overbalanced rudder used by Tabarly on *Pen Duick II*. It was found that when the lever arm of the trim tab was weighted instead of pegged no readings were possible, as the whole system was too unstable. The slightest shaking of the trim tab due to turbulence in the water was sufficient to turn the rudder by very large angles. The readings for 18 per cent and 6 per cent balance show how rapidly the sensitivity falls off.

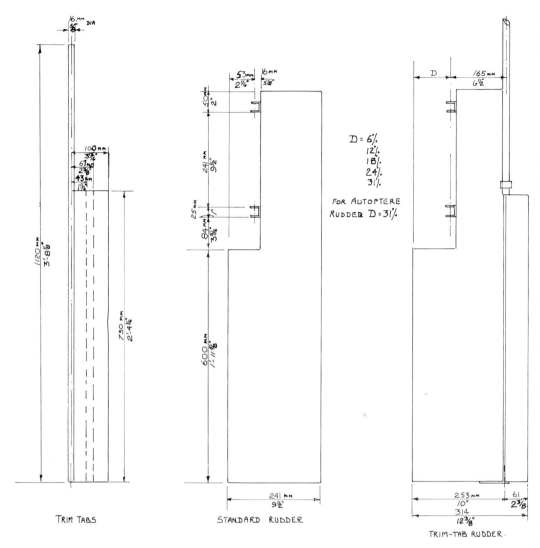

TRIM TABS

STANDARD RUDDER

D = 6%
12%
18%
24%
31%

FOR AUTOPTERE
RUDDER D = 31%

TRIM-TAB RUDDER.

Fig 80 Standard rudder and trim-tab rudders which were tested against it. The balance D was calculated using the 61mm-chord tab. With D giving 31% of balance the rudder is overbalanced and becomes an Autoptère rudder.

Fig 81 (a) Method of deflecting trim tab with weights; (b) the Autoptère rudder (see page 74) required a different set-up.

It appears that a balance of, perhaps, 22 per cent or 23 per cent, that is to say with the axis of the rudder pintles 22 per cent of the chord of the combined rudder and tab aft of the leading edge, will give the best results.

The next test was designed to find out what size to make the trim tab in relation to the overall chord. For this purpose

curves were drawn for 3 different trim tabs of 14 per cent, 22 per cent and 28 per cent of the total chord with the rudder balanced to give 6 per cent of the total chord with the 22 per cent tab. This gave the curves shown in Fig 83. This data was then redrawn to give the work required to turn the trim tabs, that is to say the weight applied to the trim tab lever

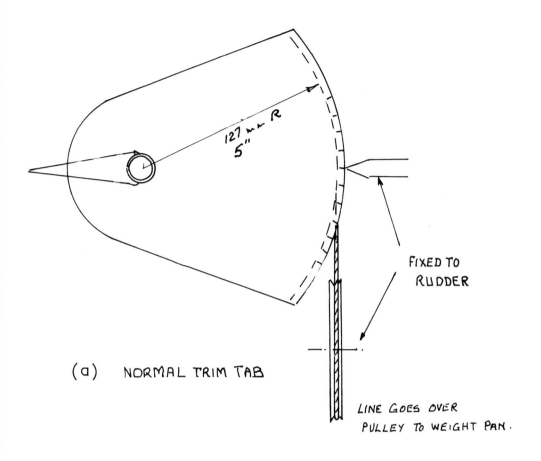

127 mm R
5"

FIXED TO
RUDDER

(a)   NORMAL TRIM TAB

LINE GOES OVER
PULLEY TO WEIGHT PAN.

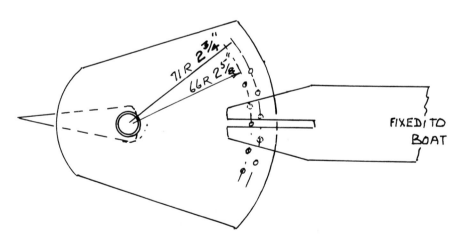

71 R 2¾"
66 R 2⅝"

FIXED TO
BOAT

(b)   TRIM TAB FOR AUTOPTERE
                RUDDER

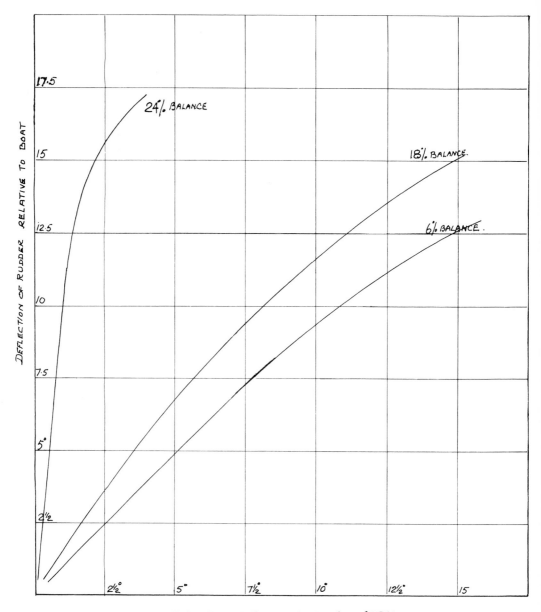

Fig 82 Deflection of trim tab: and the rudder with different amounts of rudder balance.

multiplied by the angular distance it had to move to obtain an equal angle of the standard rudder, which, in effect, measures the force the trim-tab rudder exerts to turn the boat (Fig 84).

The curves show that too large a trim tab takes too much power to turn it. The effect of the trim tab on the rudder is proportional to the area, but the force required to turn it is proportional to the area multiplied by the distance from the centre of pressure on the tab to the tab axis. Thus, the force to turn the trim tab goes up as the square of the tab chord. Too small a trim tab however, although easy to turn, is inefficient in providing enough force to turn the rudder. For a

70

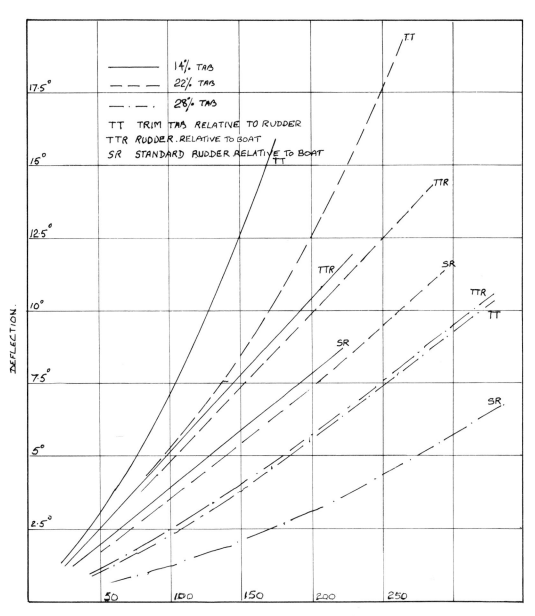

Fig 83 Deflection of rudder with torque applied to various trim tabs.

rudder which has little or no balance, a 20 per cent trim tab will give the best result. A fully balanced rudder however could probably be turned by a smaller tab. It is interesting to prove that the general rule of thumb that a 20 per cent tab is suitable seems to have been soundly based.

The next series of tests was designed to provide data on the actual behaviour of a trim tab when working with rudders with different amounts of balance. For this purpose the torque applied to the tab, which was of a standard 19.5 per cent chord, was carried well beyond the usual working limit so that the point at which the rudders stalled could be determined accurately (Fig 85).

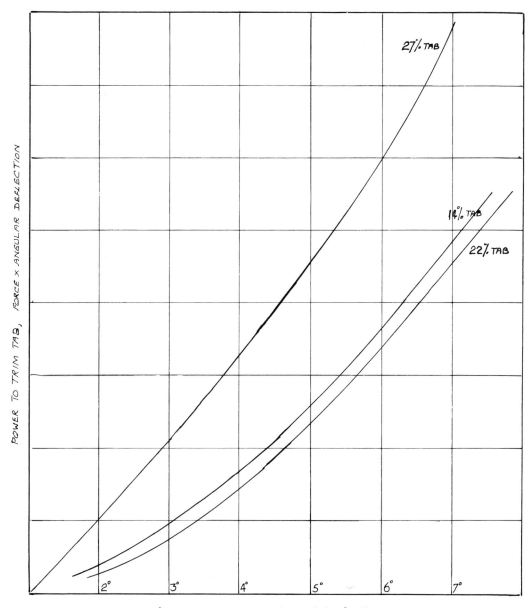

Fig 84 Power to trim tab to produce turning of boat as represented by standard rudder deflection.

It will be seen from these curves that, with a well balanced rudder, a small amount of force on the tab gives a large deflection which gives an even larger rudder deflection and this gives a large sideways thrust. This is the ideal condition. If the rudder is not balanced, as is the case when a trim tab is fitted to an existing rudder, then the trim tab will need a bit more force to turn it so as to give the same overall turning effect on the boat. Points of further significance are that the trim tab stalls at about 16° relative to the rudder and further force applied to the tab only deflects the tab without much influence on the rudder.

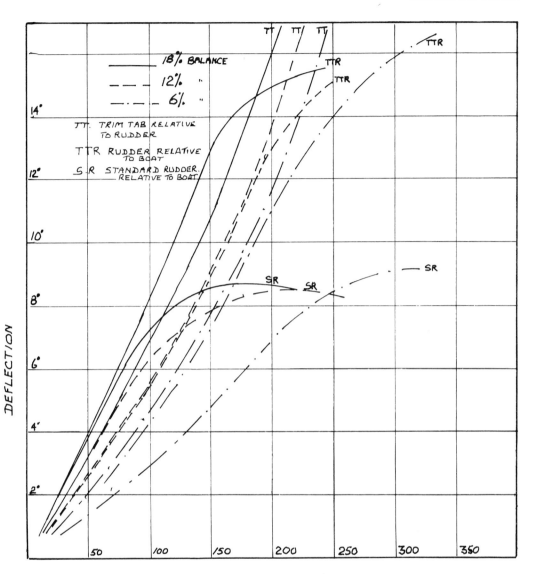

TORQUE ON TRIM TAB (GRAM CM)

Fig 85 Torque on 19.5 per cent trim tab with various degrees of rudder balance, trim-tab and rudder deflection.

The trim-tab rudder, however, stalls out at a slightly lower deflection, and movement greater than about 15° will produce no further benefit in sideways thrust.

Ideally, therefore, the vane for a balanced rudder should give a 15° movement of the tab. This will give a 15° rudder movement and a sideways thrust equivalent to about 60 per cent of the thrust to be obtained by a plain rudder of the same size, ie without tab attached, whether the rudder is balanced or not; but the unbalanced rudder will need much greater force applied to the trim tab. This, for equal performance, means that an unbalanced rudder needs a larger wind vane. Translated into practical and easily appreciated terms, a well balanced rudder can use a smaller trim tab with considerably less vane power to drive it; a rudder with little or no balance requires a slightly bigger (but not more than 20 per cent) trim tab and needs considerably more vane effort. The angles through which the tab and the rudder turn should be more or less the same, but the maximum turning effect produced on the boat in both cases will only be about 60 per cent of that produced by a plain rudder.

At small angles of attack, and particularly with a balanced rudder, the sideways thrust produced by the tab/ rudder combination may exceed that of a plain rudder, as under these conditions the tab/rudder combination acts like a simple curved foil of a larger size than the rudder itself. There is also the consideration that the vane force not only turns the tab, but also tries to turn the rudder the way it should go. Thus the total force on the rudder is that produced by the tab, plus the force used by the vane to turn the tab.

It seems that for small angles of attack, such as minor course corrections, the tab system is sufficiently powerful, particularly with an auxiliary rudder when the main rudder can be used as a trimming device so that the auxiliary only has to look after the wind variation and the main weather helm is looked after by the main rudder. Where it is necessary to couple a trim tab on an existing rudder, and the boat is not easy to control and requires the full rudder to correct weather helm, it seems that a balanced tab some little way away from the main rudder will give a better result, but at the cost of slightly greater water resistance.

The Autoptère Blade

A very interesting variation on the trim-tab idea was developed for aircraft use before the war by M. Gianoli, the designer of Tabarly's gears, and named the Autoptère. This consisted of an over-balanced rudder with trim tab, and works

as indicated in Fig 86. The steps are (a) the vane deflects the trim tab. This moves the rudder to position (b) where the trim tab is only just applied, but the water pressure on the overbalanced rudder forces the rudder further over. Stability is reached in (c) where the force on the trim tab is exactly balanced by the force on the rudder. To get a better appreciation of the working of the Autoptère rudder the reader should work out for himself what happens when the wind vane is further deflected or, alternatively, when the boat returns to its original course and the vane is again central. It will be seen that, unlike the partially balanced rudder where the forces on the tab and rudder work against each other and the net turning effect of the combination is the difference of the two, here the forces work in the same direction and the net turning effect of the rudder is the sum of the forces concerned. Not only is there this advantage, but, whereas the partially balanced rudder presents an awkward shape to the water flow, the overbalanced rudder results in a far better shape equivalent to an efficient curved airfoil (see Fig 73(a)).

A further benefit of this type of rudder is the effect it has if the boat starts to yaw due to wave or wind action (Fig 87). The Autoptère blade counters yaw not by waiting until the boat has yawed and then responding to the wind vane, but by responding to the variation of water pressure on the rudder itself. Thus there is no time lag in resisting yaw.

At first sight it would appear that such an overbalanced rudder might give difficulties when the rudder was not being used for steering, but in practice I found that, provided the trim tab was held in the central position relative to the boat, the rudder could either be used with a tiller or left to trail free.

Unfortunately, it was not possible by simple weights to measure the torque required to work the trim tab. It will be

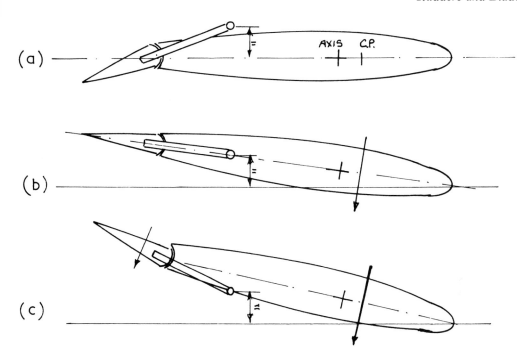

Fig 86 The working of the Autoptère rudder:
(a) the trim tab is deflected by the vane; (b) The
overbalanced rudder swings over until the
trim tab has no deflection; (c) Further
swinging of the rudder applies trim tab in the
opposite direction until force on tab is
balanced by the force on the rudder.

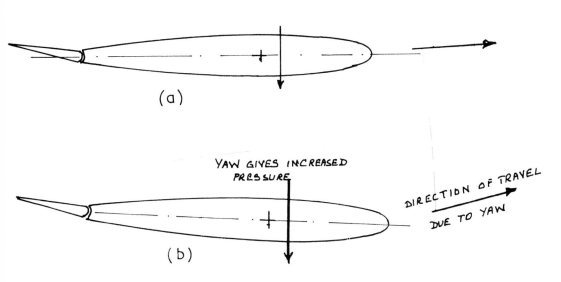

Fig 87 If an Autoptère rudder (a) is subject to
yaw of the boat the pressure on the forward
part of the rudder will increase (b) and the
rudder will turn itself to counteract the yaw.

seen from a study of Fig 86 that initially
the vane, by displacing the trim tab,
starts the rudder working; as the rudder
moves, however, it is the rudder which
changes the tab setting while the vane
just holds the trim-tab lever in position.
There is, therefore, a reversal of stresses
on the trim-tab arm which means that
simple weights could not be used. The
whole set-up was, however, extremely
sensitive.

An Autoptère blade should be made a
little stronger than the equivalent
partially balanced blade. If the boat
comes off the top of a wave and starts to
yaw, the partially balanced blade will give
way as the centre of pressure on the blade
moves aft with the blade in a stalled position (Fig 88). With the Autoptère blade
(Fig 89) the movement aft of the centre of
pressure will bring it nearer the pintle
axis and the blade will not give way.

The tests on the overbalanced rudder
went far to prove all the claims made by
the designer (for dimensions see Fig 80).
Unfortunately with this type of rudder it
is not possible to use weights to measure
the wind vane effect because, as
explained previously, there is a reversal
of forces which takes place as the wind
vane ceases to move the trim tab and the
rudder takes over. Thus the only
parameters measured were the deflection
of the trim tab relative to the boat, the
deflection of the Autoptère and standard
rudders in Fig 90 giving the result of
these trials. It will be seen that for any
angle of the Autoptère rudder the
standard rudder had to be deflected
approximately 25 per cent more to obtain
a balance. At about 17° angle of incidence
the plain rudder stalled out. The
overbalanced trim-tab rudder, however,
continued to provide turning force.

Pendulum Servo

The pendulum servo type of gear
developed by Blondie Hasler for the 1964

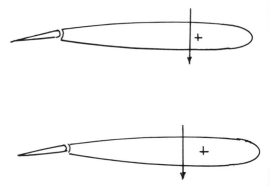

Fig 88 When a normal trim-tab rudder yaws, the centre of pressure moves aft and the rudder gives way.

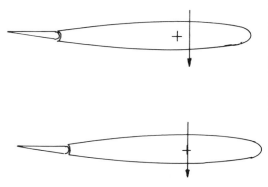

Fig 89 When an Autoptère rudder is subject to excessive yaw the centre of pressure moves nearer the rudder pintles imposing extra strain.

trans-Atlantic single-handed race was
probably the greatest single leap forward
in self-steering design as it did not depend
on the adaptation of research work done
for aeroplane control, but was an entirely
new concept.

Although the final gear may appear
complicated, the basic theory is simple
(Fig 91). If we have a long, narrow,
balanced blade in the water it is quite
easy to turn it on its axis. As soon as the
blade is turned and is at an angle to the
flow of the water, there will be a very
considerable side thrust against the
blade. This will make the whole gear
swing from side to side and this
movement can be transmitted by lines to
the boat's tiller. This gear is by far the

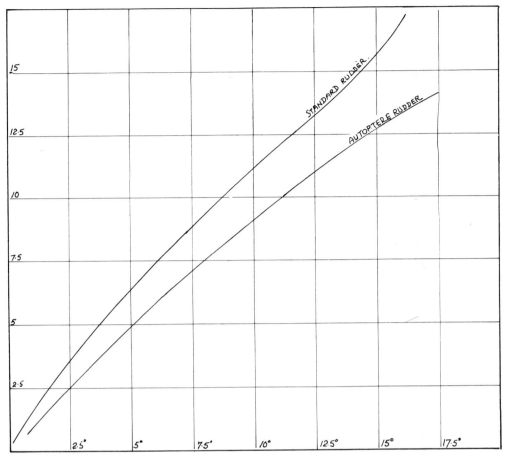

AUTOPTERE AND STANDARD RUDDER
RELATIVE TO BOAT

TRIM TAB RELATIVE TO BOAT

Fig 90 Tests on the Autoptère rudder. Note the superiority of the Autoptère over the standard rudder.

$$P \times \frac{3}{150} \text{ or } \frac{P}{50}.$$

most powerful of all gears and many commercial vane gears produced today are based on Hasler's basic idea.

It is interesting to analyse how it gets its power. Assume that the blade is held at an angle to the water flow by the wind vane with a force P applied to the blade lever arm which is 150mm long. As the blade is very well balanced, the distance from the centre of pressure to the axis of the blade is only 3mm. This, with a blade of 100mm chord, means a balance of 22 per cent which is a figure quite easy to obtain in practice. The force exerted by the vane lever arm will be

The distance from the centre of pressure of the blade to the pendulum axis is 1,200mm, and the distance of the axis from the point at which the lines are attached is 300mm. This gives a magnification of 4, or a total amplification of about 200 for the whole system. In practice, a weight of 80g applied at point P gave a pull in the tiller lines of 10kg with the boat travelling at about 4 knots. This corresponds to an actual amplification force of 120. Allowing for friction, and that the balance of the blade might not be exactly 22 per cent, the results show what can be done with this type of gear.

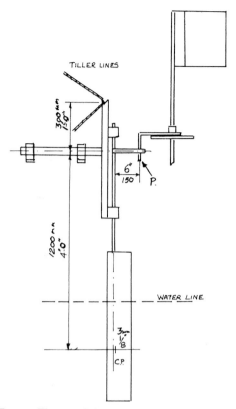

Fig 91 The pendulum servo gear is the most powerful of all types.

transmission from the vane to the blade is simple, as is the course-setting mechanism. His bearings are plain blocks of Tufnol, and his whole gear can easily be repaired by either common woodworking tools and glue, or a weld applied by the first garage met with on the boat's travels. This point was fully appreciated by me when I had a weld failure, due to a manufacturing fault, in Hasler gear on the third day out from the Canaries on a trans-Atlantic voyage. A temporary repair was effected using a bolt holding down the boat's heads, and a complete job done by a garage in Barbados which has lasted to this day having survived several knock downs and a storm which nearly tore the boat to pieces in the 1964 single-handed trans-Tasman race.

Gunning Gear

The Gunning gear (Fig 92) was the first variant on the basic Hasler idea of the pendulum servo, but had two basic differences. Gunning used a horizontal-axis vane; thus a smaller vane was possible. Also, instead of turning the whole blade, he used the blade as a skeg and turned a tab of the back of the skeg. There were several advantages to this system and, as always, some disadvantages. The tiller lines could be attached to the skeg and thus take the load off the trim-tab bearings, reducing the friction the vane had to overcome to turn the tab. By attaching the control wires further away from the pendulum axis than was convenient with the Hasler system, the control lines did not have the same strain on them, reducing the losses in the quarter blocks. The snags were that using lines from the vane to the tab restricted the vane rotation to 180°. Also the more complicated skeg and tab set-up had underwater bearings subject to fouling and, whereas the Hasler blade could easily be hinged up to clear weed, this was not possible with the Gunning.

One not inconsiderable advantage of the pendulum servo gear is the small strain it imposes on the transom of the boat. In the example given, the force on the blade is only one quarter of the force in the steering lines. Taking into account the point at which the lines are attached to the tiller and the overall length of the tiller, it can be safely assumed that the sideways force on the blade is not greater than that which an average helmsman can apply to the tiller. This force is very small compared with the force on even a half-sized rudder when a boat comes sideways off the top of a wave. On many modern lightweight boats the transom may not be stiff enough to take an auxiliary rudder, but a pendulum servo gear will impose no problems.

Hasler has always used a vertical-axis vane as this gives ample power and the exact bearings and gears. The product, although heavy by normal standards, was exceptionally well made and could not be faulted for its durability. The original in fact used a vertical-axis vane, but subsequent models were converted to horizontal axis which gives slightly better control and more power, enabling the vane size to be reduced and allowing a more compact assembly.

Today pendulum servo gears are made in many parts of the world. Details of the snags which the designer may encounter are given in the section dealing with practical gears.

With all pendulum servo gears the tiller lines are a bit of a nuisance in the cockpit and by the time there are tiller lines, course setting lines and disengaging lines the whole after end of the boat gets covered with string. This is a distinct disadvantage, but is, alas, unavoidable.

Saye Gear

An interesting hybrid which is part trim tab, part pendulum servo, is the gear invented by Saye in the USA (Fig 93).

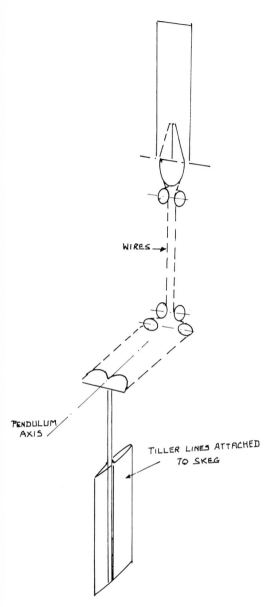

Fig 92 The Gunning pendulum servo gear used wire to transmit the vane motion.

Aries Gear

The Aries gear using a horizontal — or rather inclined to the horizontal — axis vane working a balanced blade, proved a most excellent gear. Its designer produced a far more sophisticated gear than Hasler, making use of castings instead of fabrication and using far more

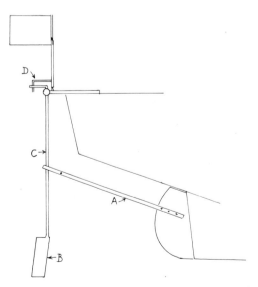

Fig 93 The Saye gear is an interesting hybrid, part pendulum servo, part trim tab.

The ship's rudder has attached to it a metal stirrup A. Through this an oar B is mounted on a metal shaft C. A vertical-axis, cloth-covered, V-shaped vane with a feed-back vane linkage D turns the shaft and oar. The oar, hinged at the top of the shaft, moves from side to side controlling the rudder. This system has many advantages: it is easy to mount on any configuration of rudder; there is considerable power leverage as, not only is the oar some way away from the rudder, but there is the advantage of the centre of effort of the water on the oar and the position of the stirrup; the oar can be retracted through the stirrup and stowed in harbour, and there are no underwater bearings to corrode or jam. The pendulum oar works like a trim tab so there is some slight loss of turning effort of the rudder as the oar works against the rudder, unlike the true pendulum servo gears where the blade turns the same way as the rudder and gives increased turning force. This system, unlike the pendulum servo, does not tend to come out of the water when the blade is trying to resist weather helm due to the boat's heeling. In the Saye gear the oar actually goes into the water under these conditions (Fig 94).

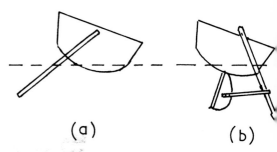

(a)   (b)

Fig 94 The true pendulum servo oar (a) tends to ride out of the water when resisting weather helm. The Saye oar (b), under the same conditions goes deeper into the water.

CHAPTER SEVEN

Potential Developments

In any development field there are ideas which seem odd ball, but which may in time be incorporated in working machines for commercial exploitation.

The Windmill

This is an attractive idea proposed by J. Morwood of the Amateur Yacht Research Society. Basically the gear consists of a windmill well geared down to drive a drum (Fig 95). It will be seen that this configuration gives what is, in effect, a lot of horizontal-axis vanes which will be turned one way or the other by the wind. Course setting is obtained by rotating the whole gear about the axis A. The force obtained in the tiller lines can be very considerable as the basic drum diameter can be quite small and not only will the leverage of the vanes be considerable, but there will be many vanes in the wind at any one time.

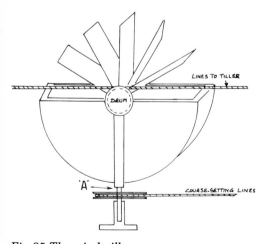

Fig 95 The windmill gear.

To give a very approximate idea of the potential efficiency of this gear, assume an overall diameter of rotor of 600mm and 12 blades, perhaps 150mm wide. The total area of blade exposed to the wind will be $150 \times 6 \times 200$ (this allows for the space taken up by hub and winch) and amounts to 0.18sq m. The lever-arm distance of the vane to the centre of the hub is 150mm, so that the total effect is proportional to $0.18 \times 0.15 \times 150$, or 0.027^3. A 1,200mm $\times$ 300mm vane has a leverage of 0.216^3, or 8 times the effect of our windmill. The vane can, however, turn about 45° on each side, whereas the windmill might be made to give 3 turns each way. This would result in an overall superiority of 2 to 1 in favour of the windmill. The only snag might be the slow response. A further difficulty could be the complete absence of feed-back which would probably give very erratic steering.

The Gyroscopic Stabilizer

The thinking behind this control is that if the boat yaws, or suddenly accelerates down a wave, the normal wind vane may for a time lose control and some system of stabilizing the boat's steering is needed until the vane again becomes effective. The gyroscopic system has, as far as I am aware, only been tried out in practice by M. Gianoli on his MNOP64 gear. The idea was that the wind-driven gyroscope should counteract the effect of yaw by turning a flap on the trailing edge of the wind vane. The attempt was not altogether satisfactory, possibly because

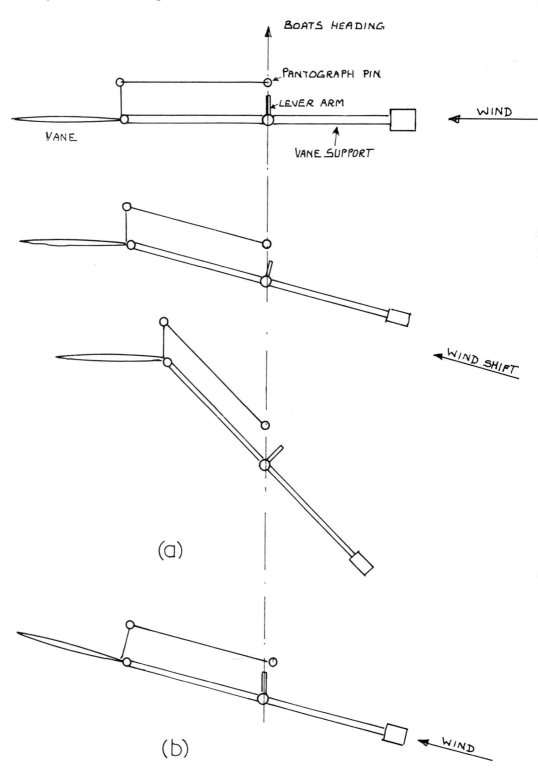

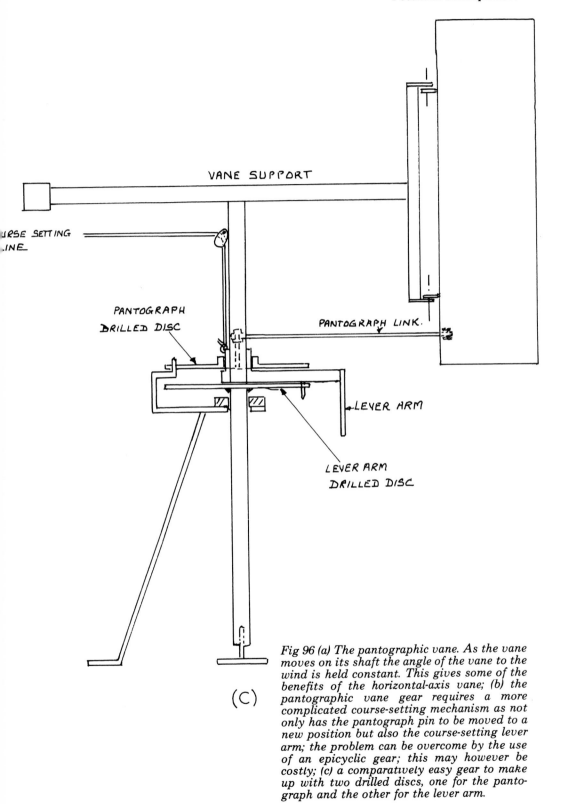

VANE SUPPORT

URSE SETTING
LINE

PANTOGRAPH
DRILLED DISC

PANTOGRAPH LINK.

LEVER ARM

LEVER ARM
DRILLED DISC

(C)

Fig 96 (a) The pantographic vane. As the vane
moves on its shaft the angle of the vane to the
wind is held constant. This gives some of the
benefits of the horizontal-axis vane; (b) the
pantographic vane gear requires a more
complicated course-setting mechanism as not
only has the pantograph pin to be moved to a
new position but also the course-setting lever
arm; the problem can be overcome by the use
of an epicyclic gear; this may however be
costly; (c) a comparatively easy gear to make
up with two drilled discs, one for the pantograph and the other for the lever arm.

the lively motion of any light, fast-moving boat would tend to confuse any gyroscope. When Gianoli produced his next commercial vane, the MNOP66, he abandoned the use of the gyroscope and used instead his Autoptère rudder which was simpler and far more efficient in counteracting the effect of yaw.

The Pantographic Vane Linkage

This idea, developed by Derek Fawcett, seems an attractive proposition, combining as it does the advantages of the vertically pivoted vane with its simple linkage from vane to rudder with the benefits of a vane which does not change its angle to the wind as it moves.

Fig 96a shows the layout of the gear. When there is a wind shift the vane support will rotate, but the angle of the vane will be held constant to the wind by the pantograph mechanism. The vane-support rotation will continue until the force required to work the steering mechanism is balanced by the force of the wind on the vane less the effect of drag on the vane which works against the vane efficiency. Derek Fawcett claimed that the vane support could rotate up to 70° for a 5° wind shift, but this claim seems excessive. There is no reason, however, why a rotation of 45° should not be possible and effective; thus a vane of this type should be, area for area, fully as effective as a horizontally pivoted vane. By altering the length of the pantograph arms a further increase in efficiency could be obtained, as the vane could be made not only to hold its angle to the wind but actually to increase this angle as the vane-support rotated.

Altering course is rather more complicated than with a simple vane (Fig 96b). Not only has the position of the lever arm to be altered relative to the vane support, but the position of the pantograph pin has to be altered relative to the boat.

Derek Fawcett used a moving-carriage gear chain to obtain this result, but the gears are rather sophisticated and beyond the capacity of the home workshop. Fig 96c gives a simpler mechanism which is not too hard to make up. The gear consists of a vane hinged on the vane support which is welded to a vertical tube similar to that used on a vertical-axis vane. Welded to this vertical tube is a course-setting disc with a lever arm. Free to rotate on a spigot, which is part of the lever arm, is a second drilled disc on which is fixed the pantograph pin. This disc can drop onto a pin attached to the boat.

To set course the course-setting line is pulled and the boat hand-steered onto the course desired. As both the vane support and the pantograph discs are free, the vane support will rotate until the vane is facing into the wind with the pantograph mechanism also moving into the neutral position. Releasing the course-setting line will fix the lever arm relative to the vane support and also the pantograph pin relative to the boat.

When this method is compared with the horizontal-axis vane there appears to be little advantage. Size for size, there will be little difference in the efficiency, but it appears that the course-setting mechanism of the Fawcett gear will be more complicated and, thus, more expensive, and the vane will require a larger area free of obstruction than is required by the horizontal-axis vane.

Epicyclic Vane Setting

An epicyclic gear chain to turn a vane for course setting has been proposed and used successfully by Albert Wilcock on model yachts. This system is usually referred to as the moving-carriage mechanism and Fig 97 shows how it works.

The gear A is attached to the rudder, and the gear B to the wind vane. By rotating C (the moving carriage) the angle

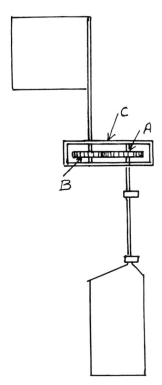

Fig 97 The moving-carriage gear. This uses an epicyclic train where the rotation of the carriage produces double the rotation of the vane shaft. This can give stepless adjustment of course simply by rotating the carriage.

of the wind vane attached to B is altered. A 45° movement of the arm C gives 90° movement of the vane on its axis. While the carriage C is held in one position the vane turns the rudder just as in any simple gear mechanism. It is assumed in this example that the two gear wheels are of the same size. This system has not been used commercially on full-sized boats probably due to the expense of production, and the course setting does not appear to be superior to the simpler worm and latch principle.

Transmission of vane movement to rudder by means of bowden cables or hydraulics could, under certain circumstances, provide the answer to an otherwise insoluble problem, as either system can be used to link the vane to the rest of the gear round corners and over considerable distances. The question of friction could arise with bowden cables, but some successful gears have used these; presumably it is a question of good design. Hydraulic transmission could give very good results, but here again friction in the pistons might be excessive, and while use of capsules instead of pistons would reduce the friction it might not be possible to get sufficient movement to do the job. It is possible, however, that a hydraulic coupling from the vane could control a master unit which would then do the steering. This would, however, probably be expensive unless standard units could be incorporated in the design and the boat already had hydraulic steering. The vane could then be used to control the valve system.

CHAPTER EIGHT

High-speed Sailing

When multihulls are sailing in brisk winds they can behave in ways which baffle the normal vane operated self-steering gear. The phenomena are referred to as multihull breakaway and downwind surfing.

High-speed breakaway can best be described by the set of examples in Fig 98. Assume a true wind strength of 10 knots and we are close hauled doing 4 knots. We are beating into the waves which tends to slow us up and the position is fairly stable. If, now, the boat strays off course, or a wave turns it, the boat will pick up speed. Although the relative wind remains in the same direction, the boat will be heading off the true wind. The effect of the waves slowing up the boat will now be reduced and the boat may again accelerate. This will again result in the boat going further downwind in order to keep the relative wind in the same direction. The last stage of all is with the boat careering wildly across wind with the waves helping rather than hindering, but with the relative wind still in the same direction and the self steering still thinking it is going close hauled to windward!

In Fig 98 the boat's speed varies from 4 knots to 16 knots, whereas the relative wind speed, the wind over the deck, only changes from 13 to 19 knots. Although we have a four-fold increase in velocity through the water, the increase of air velocity is only about 50 per cent. Thus, it would seem that any control to stop the multihull breakaway should be powered by water velocity, not air.

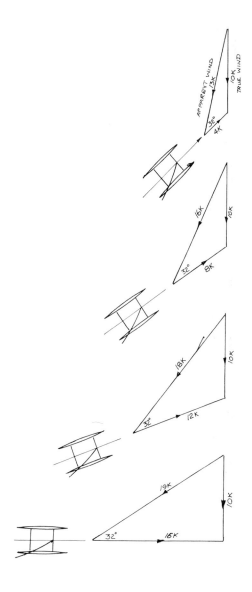

Fig 98 Vector diagrams for high-speed breakaway showing how the apparent wind direction can remain constant, but the boat speed and heading can vary enormously.

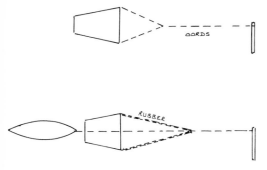

Fig 99 *A drogue can be made to luff a boat to windward when the speed becomes excessive, by acting on the tiller or wind vane.*

To reduce the boat's speed there are two courses open: use brakes similar in function to the dive brakes of an aircraft; or use a control which gives an automatic luffing action as the boat's speed builds up. This will reduce the effect of the sails and lower the speed. So far proposals have been made to stream a drogue which can either just be a sea anchor, or which

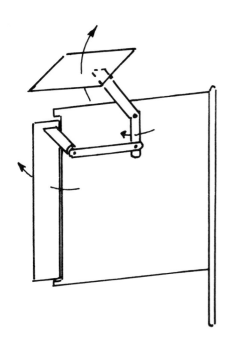

Fig 100 *An alternative is to have a tab on the wind vane which is deflected when the wind speed rises.*

can be connected to the wind vane as in Fig 99, or to have a wind vane where increasing air velocity operates a trim tab on the vane changing its setting as in Fig 100. We are apparently compelled to have this extra drag all the time, so that as soon as we reach a dangerous speed it will work. The snag could well be that not only is there the extra resistance but, as anyone who has tried it will know, a drogue in a rough sea is a most fickle animal and would, in my opinion, be dangerous anywhere near a boat's steering. The wind vane/wind velocity sensor would seem to be more suitable if it were not for the only comparatively small increase of apparent wind with breakaway conditions.

Rubbing the slate clean and going back to first principles we require something that, with minimum water resistance, can provide a controllable force which can be used to luff the boat into wind until such a time as the boat's speed has dropped. Hasler's pendulum servo mechanism seems to be the ideal device for the job.

Imagine a very small oar held in the water at a slight angle. This exerts a considerable sideways force, but the drag is only about $\frac{1}{40}$ of this force. If we now use our oar on a pendulum we can pre-load it so that it only moves when the speed rises above a preset value. The movement of the pendulum can then be used to act on the vane or self-steering linkage to give the effect we want (Fig 101). The device consists of:

A A small hydrodynamically balanced oar in the water.

B Bearings on which the oar works as a pendulum. These bearings are housed in such a way that the housing can rotate a small amount in each direction.

C A lever arm.

D A spring.

E Stops which hold the pendulum vertical against the spring.

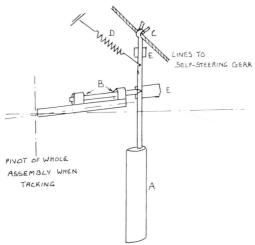

Fig 101 A proposal to eliminate high-speed breakaway by the use of a small pendulum servo oar.

In action the spring holds the pendulum against the stop and also turns the bearings B and thus the vane A, at a small angle to the water flow. Nothing will happen until the water flow attacking the oar at the preset angle overrides the spring and moves the pendulum. This movement is then transmitted to the vane or other part of the self-steering mechanism. On the other tack the spring is moved to the other side of the pendulum. This gear gives the minimum water resistance for the maximum force produced. The only disadvantage is, possibly, complexity, and the steering action of the oar on the boat. This steering action will be small and will be easily corrected by the main rudder without any appreciable increase in drag.

The question of high-speed surge down a wave front is met with on keelers, but is particularly common with multihulls where phenomenal acceleration and ultimate speeds are common when riding downwind with large waves. Under these conditions the boat can actually travel down a wave front at a speed greater than the wind, setting the sails aback and completely flummoxing the wind-vane self-steering. The vectors which have changed are a rapid increase in speed and an equally rapid fall in wind velocity over the deck. There is no real cure except, perhaps, dive brakes; but one thing which must at all costs be avoided is a change of course as we go surfing down the wave front which will probably result in a broach and potential capsize. It seems nearly always to be waves rather than wind which turn boats over.

The best we can do is to fix the rudder as near as possible in the straight running position and hold it there. The question we should ask is what mechanism, under falling wind velocity, will tend to hold the rudder so that it has no tendency to turn the boat one way or the other? The answer is simple. Under surfing conditions when the true wind velocity will be plenty to work a wind vane, just overweight the balance weight of a horizontal-axis wind vane. When the boat is not surfing there will be enough wind to work even an overweighted vane. When the boat starts to run downhill the wind will drop, but the vane weight will centralize the vane and thence the rudder. A similar effect would be obtained by shock cord each side of the wind vane of a vertical-axis vane. When the wind drops, the lever arm will automatically centralize and centralize the trim tab and the servo rudder it controls.

The solution to the problem assumes a reasonable sail balance. If the sail plan is ill-balanced and the boat has a natural tendency to screw up on the run, it is obviously no use just to centralize the controls. As always, sail balance should be the first requirement of good seamanship. Just because the boat is fitted with self-steering is no excuse for bad sail choice or setting in the hope that 'George' will override any of our own failings.

As a possibly bigoted keeler sailor it does seem to me that prudent sailing is the answer, rather than gimmicks which are added to the boat and which will most certainly provide a boarding ladder for Mr Murphy to come aboard.

CHAPTER NINE

Types of Boat and the best Gears for them

Tiller steered boats are by far the easiest to fit as the rudder stock usually moves freely. If the rudder is stiff then, instead of the self-steering continually applying small movements to a rudder in a constant state of motion, the rudder will remain fixed until the gear develops enough power by the boat going off course to move it. The rudder will stay in the new position until it can be jerked back again by a considerable swing in the other direction. Before going further every attempt should be made to loosen the rudder. If this is not possible then a powerful gear will be required, probably a pendulum servo which gives by far the most force for a small course alteration.

If the tiller is free and the boat small and easy to steer, then a plain horizontal-axis vane coupled to the tiller by lines will do the job quite well and is an easy and cheap gear to make. How large a boat this will suit depends very much on the individual boat and its rudder. My present boat *Tabitha*, hard mouthed, of 7.5m, steers quite well with a horizontal-axis vane until reaching in about 15 knots of wind. At this wind speed the boat is leaning over and the weather helm then becomes considerable. Trimming the sails by reducing the mainsail counteracts this, but finally the boat's effective speed has to be reduced. In effect, as the boat becomes harder pressed, it develops weather helm, and when this becomes considerable sail has to be reduced if the self-steering is to cope. For day sailing where the self-steering is just to let me be lazy or do other jobs this limitation does not matter; if however I was trying to race the boat single-handed, having to reduce sail prematurely would be a disadvantage. Also I can never go to sleep without the certainty that, whatever the rise or fall of the wind strength, my boat will continue to sail the course set.

The boat in which I and my wife sailed the Atlantic was a 24ft Guy Thompson designed T24 with a separate skeg and rudder configuration. This, because of the load taken off the rudder by the skeg, was easy to steer and it was well-shaped and did not develop much weather helm until after the gunnel was awash. Guy Thompson designs his boats on the basis of sailing models. These, unless they are sweet to sail, will not sail by themselves and, like the models, the finished boats are a pleasure to steer. This boat, with a horizontal-axis vane coupled to the tiller by lines, could be sailed and controlled under winds up to 35 knots.

Josephine, originally 10.6m until I cut 0.6m off her sharklike bow, was designed to steer herself and the rudder was balanced. A horizontal-axis vane gave complete control under winds of up to 45 knots at which speed I suffered knock-downs which made the cabin a bit untidy, but the tell-tale compass in the cabin indicated that there was no need to open the hatch and re-set the steering. A trimaran of 11m long gave good control, but I never personally tried to drive it in a storm. A Wharram catamaran of 14m overall also behaved quite well with a simple horizontal-axis vane directly coupled to the tiller for most of the time.

From this it will be seen that a normal keeler of about 9m maximum, or a trimaran of up to 12m can be steered by a simple horizontal-axis vane if your requirements are not too severe and the vane is big enough. If closer steering, or ability to sail on in stronger winds is required, a balanced rudder over the stern controlled by your wind vane will give all the power required. I have personally sailed a number of ocean cruisers back from overseas races of over 1,000 miles each using an auxiliary rudder of this nature, and the watch keeping on the return journey was only a very small fraction of the strain that had gone into the race itself.

A pendulum servo gear will produce as much force as a single helmsman is capable of exerting, so this type of gear can be used to sail a tiller steered boat of any size. The extra complexity of this type of gear may give the mildly enthusiastic something to play with, but it is not easy for the amateur to build and so, unless time and inclination are there, may be better purchased.

Where the boat's main rudder sticks out at the back an auxiliary rudder is hard to fit, although two balanced rudders connected together as in Fig 102 give a good result. Alternatively, a trim tab can be fitted to the main rudder. The 1978 single-handed trans-Tasman race was won by a boat using a gear of this type designed by myself. Winds of at least 50 knots were encountered and two boats out of a fleet of fifteen were lost, so it was a good test. The vane, for simplicity, can have a vertical axis; and vane and trim tab can be of any size to suit the rudder. The success of this type of gear depends on how easy the rudder is to turn. A rudder mounted on a skeg usually gives no trouble, nor does a partially balanced spade rudder. Some of the 'barn doors' may require more power than a trim tab can supply.

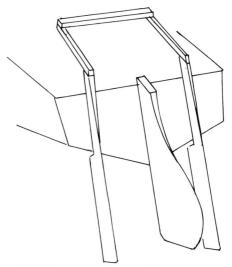

Fig 102 Twin auxiliary rudders can be the answer to a boat with a rudder on the stern.

When it comes to wheel steered boats these usually have the rudder tucked in well away from the transom and this creates problems. Pendulum servo gears have difficulty in turning the main wheel, which needs turning up to a full turn each way. This, on a 150mm diameter pulley, means 450mm of travel for a piece of string wound round it. As the pendulum servo gear only travels about 125mm each side of the vertical, allowing a bit for rope stretch, we need about a 4:1 ratio. This can be achieved by a double fool's purchase (Fig 103).

We can, perhaps, use only one purchase if the pulley is attached to the ship's wheel by means of a pin. The pin is withdrawn and the wheel set to steer the boat on its correct course; the pulley is then fixed to the wheel. By this method the pendulum servo only has to look after the course or wind variation. The main weather helm is trimmed out before the self-steering is coupled in; thus a half turn each way should control the boat. When you take into consideration the complexity of lines and purchases, also the probable distance of the servo gear to the wheel position, the pendulum servo will do the job, but at the cost of consider-

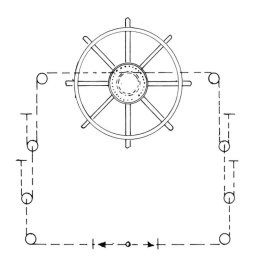

Fig 103 A fool's purchase to give extra rotation of the wheel with only small movement of the pendulum servo gear. This works, but a lot of power is lost.

working the vane latch gear. To set course, the latch gear is disengaged. The boat is then brought onto a course slightly below that which it is desired to sail. The wheel is then fixed and the latch engaged. The boat will then luff up a little, but as soon as there is enough wind on the vane the gear will take over and steer the boat. In practice, it is better not to trim out all the weather helm by the main rudder as, if the wind suddenly drops, the main rudder will overpower the self-steering rudder. The exact balance of the fixed load to be carried by the main rudder and the load variations carried by the self-steering can only be found in practice. The larger the auxiliary rudder, the less trim is required from the main rudder; it is usually found that an auxiliary rudder about half the area of the main works well enough. I have no personal experience of the Saye rig (see page 79), but this does seem to be a good solution, providing a compact outfit which can control rudders even when they are well tucked in under the boat. Using a trim tab on a tucked-under main rudder which requires all sorts of underwater mechanism is not recommended as it will either corrode, foul, give a lot of water resistance, or all three. I have designed auxiliary rudder trim-tab gears for boats from 9m to nearly 18m overall, and although under extreme conditions the control may not be quite as good as that exerted by a pendulum servo, it is still very good.

able inconvenience. If the boat has an emergency tiller steering set-up, and it should have, then the pendulum servo can work the emergency tiller without trouble. To get the best results the main steering cables from the wheel to the rudder should be disconnected. This will reduce the friction and make the steering lighter and more responsive.

For normal use, however, an auxiliary rudder set-up, either a trim tab and vertical-axis vane, or a fully balanced rudder with a horizontal-axis vane, will work well. The trim-tab set-up is particularly recommended as the only control which is necessary is a simple line

CHAPTER TEN

Some Difficult Jobs for Self-steering

The most common difficulty to be met with in all boats is very stiff steering. This may be due to a number of causes, some of which can be cured. Wheel steered boats may have too much friction for good steering. Slackening the steering cables may reduce the friction, or the wheel cables may be disconnected completely and the emergency steering tiller used with the self-steering coupled to it. Where the rudder itself is stiff it will probably require a major operation to loosen it up. In this case an independent rudder and trim tab, or an independent balanced rudder is probably the answer.

Before anyone shoots me down in flames I should, perhaps, emphasize that anyone knowledgeable of the workings of self-steering gears could probably make almost any gear work any boat if the conditions were right. Conversely, I have met the most hopeless mess made by yachtsmen who did not know what they were doing. All my remarks are aimed at the average yachtsman who wants reasonable performance under most conditions. Gales and flat calms always cause troubles and not only to self-steering gears.

The pendulum servo gears usually have the following lines going from the gear to the steering position: two lines to work the tiller or wheel; two lines to set the course; one line for disengagement. All this string all over the boat may get in the way when the steering position is amidships. An independent rudder and trim tab worked by a vertical vane only requires one line to work the latch; a balanced

rudder and horizontal-axis vane requires two lines to turn the vane for course setting.

Apart from the steering position, the most usual problem is due to the effect of the mizzen sail, particularly in yawls when the mizzen boom projects over the stern. One solution is the low vane. This is inefficient, partly because the vane may be in the wind shadow of obstructions on the deck. The size of the vane is also limited, and on close hauled courses will not be in the clear wind but in the downwash of the wind off the sails. The vane will never be in clear, undisturbed air (Fig 104).

I have met two yachts which cured the problem. Both required a little fixing as the boat came about on a tack, but both sailed well and ocean voyages proved the reliability of the steering. A little trouble with tacking is not serious, as the number of tacks made by an ocean cruiser is not large.

Moriah, an American yacht, had an auxiliary, semi-balanced rudder worked by a trim tab set on the stern (Fig 105). On each quarter was a position to house a vertical-axis vane. The gear consisted of a vertical-axis vane with a course-setting disc low down on its shaft, with a rod connecting this disc to the trim-tab lever. Stops were provided on the connecting rod to prevent the linkage travelling over dead centre. Care should always be taken with any form of linkage to make certain that it cannot travel over dead centre as the results can be disastrous.

To tack, the ship required the following

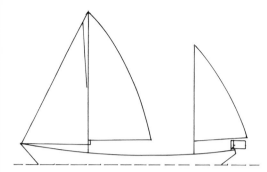

Fig 104 Boats with mizzen sails projecting over the stern are difficult to fit with self-steering.

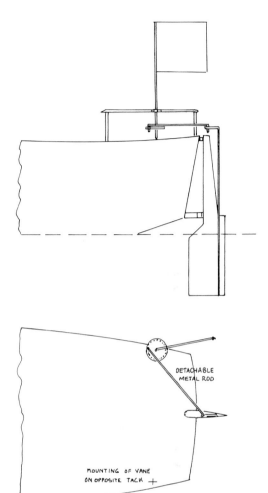

Fig 105 Arrangement of self-steering on Moriah.

procedure. Remove the connecting rod and dismount the vane — a simple procedure as the lower bearing was a simple pin bearing and the upper bearing was held to the pushpit rail by means of a quick release clamp. The ship was then tacked manually and the vane and connecting rod set up on the other tack. There should always be a preventer on the mizzen boom to prevent an accidental gybe. When running the boat used a twin headsail rig and the mizzen was not required.

For a delivery trip from Auckland to Suva, a distance of about 1,100 miles, the ketch *Portofino* was fitted with a balanced auxiliary rudder on the stern (Fig 106). This was worked by either of two horizontal-axis vanes mounted on the quarters. The transmission from the vanes to the auxiliary rudder tiller was by means of lines.

In use the leeward vane was tied down to the handrail and the windward vane only was used. When the boat had to be tacked, both vanes were disconnected and tied down; the boat was then tacked by hand. The tack completed and the boom preventer secured, the windward vane was united from the handrail and coupled to the auxiliary rudder tiller. Both vane tiller lines were permanently rigged, but only the lines in use were coupled to the tiller by means of a chain in the lines and a pin on the tiller. We thought that, if

running, we could drop the mizzen and set larger headsails using both vanes to get greater steering power with the reduced wind available. On trial this seemed to work, but we never had the wind astern for any length of time when the actual delivery was made.

This gear was, perhaps, a little over elaborate with its two vanes, but as they were only made of plywood and glue the cost was very little and, in practice, tacking was no difficulty. The whole gear worked well. During the night, watches were mounted; but in the daytime, while

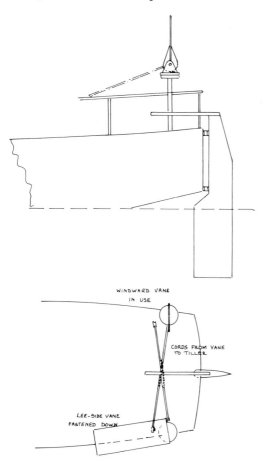

Fig 106 Arrangement of self-steering on Portofino.

On some boats, mounting an auxiliary rudder is not easy. This is particularly so with rather old-fashioned boats having full bodies and small transoms. There are two disadvantages. The small transom is not large enough to mount an auxiliary rudder securely and the shape of the boat has a tendency to suck the water up the stern. If it is intended to mount an auxiliary rudder on such a boat it is as well to go out sailing in it to find out where the water line comes when in motion (Fig 107). This will avoid the mistake I made when I fitted a pendulum servo gear to a Herreschoff H28 without sailing in the boat. Rounding the East Cape in the Round the North Island race in New Zealand, with a very nice gale behind and the clock registering 9 knots, the self-steering gear was practically under water.

There is one other case where a trick is used to get better and more stable steering. On the reach or beat the wind presses the sails and this keeps the boat from rolling; on the run however with the wind dead aft the boat may start rolling and this rolling, because of the shape of the hull, results in the boat going off course. The wind vane has a job to control this course deviation, because with the wind dead aft the vane is pointing directly at the wind and is in a dead spot. That is to

the majority of the crew were studying navigation, Keri, aged eight, who was usually playing about somewhere on deck, kept a look-out and a watch on the fishing line. We ended up probably the most over-navigated boat to leave New Zealand, so much so that, approaching Kandavu Island outside Suva at night after playing about with stars and moon, we were certain the Washington light was not burning. Dawn showed the unmistakable end of Kandavu Island, but no light, as apparently it had not recovered from a hurricane the month before. As well as relieving the crew, self-steering results in far better navigation. After you have read a book, caught a fish and had lunch, there is nothing else to do.

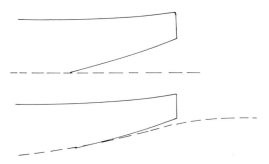

Fig 107 Where boats have long counters it is advisable before fitting self-steering to find where the water comes when the boat is sailing fast.

say, no steering is done by the wind vane and it requires an appreciable deviation to port or starboard of the boat before the wind is at enough angle to the vane to apply the rudder force necessary to control the wandering.

The cure for this is to trim the sails so that the boat only wants to go one way, or to load the tiller with shock cord. This means that the self-steering only has to fight on one direction, and any tendency of the boat to wander will be met by an immediate increase or decrease of the vane force without trouble from the dead spot when the wind is dead aft.

PART TWO

Design and Construction

Introductory note

The drawings which follow are all gears which any handyman of reasonable competency — and I include myself in this category — can make up. Nearly all my original designs have been made in wood as this material is a lot cheaper than stainless steel; and my total tool chest does not contain much more than that used by any carpenter. The only exceptions are a hand-held electric planer and an electric hand-held circular saw. These two are not strictly necessary, but they do save a lot of time and energy.

Where I have shown wood as the material, a man with access to welding and metalworking machines can fairly easily redesign to suit his materials. This is the process I myself follow. My original wooden gears, if proved satisfactory on a boat, are redesigned for manufacture in metal and the customer then gets a beautiful, shiny strong gear; but the snag is that, unlike most development work, the production version costs a lot more to make than the original prototype. The keen-minded might well ask, 'Why not put the prototype on sale?', which prompts the further question, 'Who wants a self-steering gear made of plywood and glue?' In fact it is all right for the man who gets a lot of fun out of making and using his own gear and is prepared to put up with something a little more husky and not so smart to look at. It won't perhaps last as long, but anyone who can make the gear can certainly repair it if it goes wrong, even when the boat is in the middle of the Atlantic or Pacific. Few of us carry welding gear with us, but some spare bunk-boards and glue will work wonders on a plywood gear. Anyway, it seems a little inconsistent to sail a plywood boat then say the self-steering must be of metal.

Certain parts I have shown as made in metal. Here stainless is the best metal to use, but steel fabricated and heavily galvanized will give good service even if it does not look so smart. The one thing to avoid at all costs is to mix metals, as immersion of dissimilar metals in salt water results in the rapid destruction of one or other of them due to galvanic action.

The examples can be copied exactly if they are of the correct size, but are chiefly given so that the details governing design can be pointed out thus allowing the reader to do his own designing based on sound principles. As an example of how successful this can be, I can only quote an instance where I myself was fooled. For the 1974 trans-Tasman solo race we were going to have one female competitor and we were all interested to see what she and her boat looked like. The boat turned out to be a large, lumbering, ferro-cement vessel — quite unlike the owner, who was small, slim and mini-skirted. However in spite of her charms I was more interested in her self-steering gear which was a pendulum servo, obviously homemade, but in which all the design points which really mattered had been considered and a really workmanlike job achieved. After formal introductions were over I complimented Annette Wilde on her gear and the thought which had obviously gone into it. 'Don't you know where we got the idea?' she asked. I denied any knowledge. 'Off your own drawings which you made for the magazine *Sea Spray*!' That set of drawings had shown a pendulum servo gear worked by a vertical-axis vane all made from plywood and dinghy fittings — the same materials, but a very different set up to that shown in this book. Annette and a friend had, step by step, analysed my drawings and then redesigned the whole thing in metal, but still keeping to the same general dimensions and principles. The result worked and took her to Australia over a very stormy Tasman Sea.

These drawings are to enable you to make your own gears without having the disappointment of finding at the end of a lot of hard work and not a little expense that the gear does not work.

In general the designs are as simple and uncomplicated as it is possible to make them and still get a reasonable performance. In my early days in Africa we used to say that any machine that was a deviation from a solid block of iron was a mistake. Exposed to all the known and unknown stresses which the sea can impose on a small boat, simplicity of construction is in my opinion more important than over-enthusiastic complexity in order to obtain some slight theoretical improvement in performance. The sea will always find out weak points and the more complicated the gear the more weak points there will be. A complicated gear may look nice and it can have each detail described in glowing terms in the advertising pamphlet, but when the boat has been rolled in a gale it is the simple gear which can be repaired. *Josephine* did two 360° rolls with successive waves before landing up on Middleton Reef. The OGT I vane gear was broken away from the pushpit but ended up intact, floating beside the wreck, still held to the boat by the tiller lines. I finished with an unserviceable boat but with a self-steering gear which I could have easily lashed back into position with some string!

CHAPTER ELEVEN

Wind Vanes

Vertically Pivoted

Solid vanes, that is to say those made of plywood or other similar material, should be approximately of the shape shown in Fig 28. From one 1,200mm × 2,400mm sheet of 4mm thick plywood, four wind vanes can be cut (Fig 29). The top end can be rounded and the bottom corner can be cut at an angle to miss the backstay. The forward edge can have standard beading glued to each side; 40mm × 12mm wooden strips each side of the vane hold it to the vane shaft. The top arms are extended to take the balance weight. The vane can be made of 4mm-thick plywood as this is quite strong enough. Never forget that this vane only gets turned by the wind, so must be as light as possible. If the vane flexes slightly this does not matter; it is like the wing of an aeroplane — if it did not flex it might break. There should, ideally, be a gap between the vane and the vane shaft of about one quarter of the vane chord. The vane shaft should be of tube of between 25mm and 32mm outside diameter.

The vane will need two bearings. The top one can be Tufnol, Teflon or, for economy, hardwood soaked in oil (Fig 108).

It must have plenty of clearance as the motion of the boat will let the vane rattle into position; too tight a bearing is useless. The bottom bearing can also be a Tufnol, Teflon or hardwood plug driven into the tube and pinned there. The plug

Fig 108 Top and bottom bearing of a vertical vane can be made of Teflon, Tufnol, or wood soaked in oil.

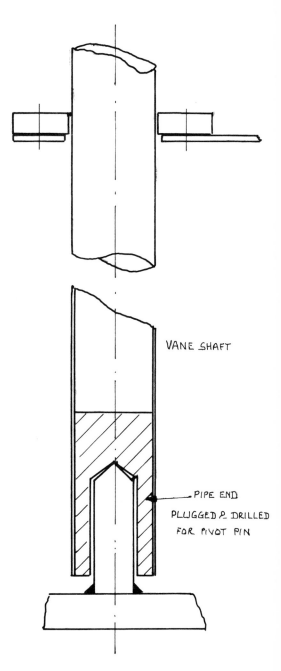

VANE SHAFT

PIPE END PLUGGED & DRILLED FOR PIVOT PIN

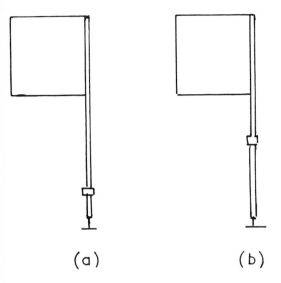

(a)   (b)

Fig 109 The top bearing should not be low as in (a) but as high up the vane shaft as is convenient (b), as rough seas will impose very heavy loads on the shaft.

is then drilled to take a pin fixed to the steering-gear framework. This gives a pin-bearing with surprisingly low friction.

Sizes of Vane

I have found that a vane cut from a quarter of a sheet of ply as described above can be used on the majority of gears using a vertical vane. When a pendulum servo system has a blade of 100mm to 150mm chord, a smaller vane can, perhaps, be used. When dealing with the larger trim-tab rudders I have increased the size to approximately 1,650mm × 700mm, but it was only necessary on boats of over 14m and up to 18m long. I must confess I have never fitted a gear to a boat over 18m so cannot, with certainty, say what is required.

Adjustments to Vane

The counterweight should just balance the vane about its shaft so that, as the boat heels, there is no tendency for the vane to give a false wind.

Care should be taken that the top bearing of the vane is as high up as possible to reduce leverage of the weight of the vane which, in a rough sea, can set up considerable stresses in the shaft (Fig 109). If the vane does not appear to do its job properly then ears on the back of the vane (Fig 30) will increase its effectiveness, but watch the vane shaft.

V-shaped Vanes

These have certain advantages. They are light, easy to dismantle in harbour and easy to make; the only snag is the extra load in the vane shaft and bearings due to the vastly increased drag. But these advantages only apply when the vane is of cloth over a frame — a V-shaped plywood vane is heavy without corresponding benefit in performance, and is not recommended.

A simple such vane is shown in Fig 36a. The vane shaft has four arms welded on to it and the cloth, with eyelets at the corners, is lashed to loops on the arms. This is probably the simplest and lightest form of vane.

An alternative which gives a slightly stiffer, but rather heavier, vane is to make the framework as shown in Fig 36b and lash the cloth to this. The exact size of the shaft and framework is not critical and any available material can be used. The cloth should be stretchless material such as Dacron or Terylene. The actual size of the vane can be slightly smaller than the flat-plate equivalent.

Horizontally Pivoted

As mentioned earlier, the horizontal-axis vane was originally used by Gianoli for the gears he developed for Eric Tabarly's *Pen Duick III* and, as soon as it became obvious that he had discovered a far more powerful vane than the vertical-axis gears, other designers were not slow to develop his ideas. But whereas Gianoli gears were highly sophisticated, those developed by the AYRS were designed for cheap home manufacture. The vanes developed by myself owe their origin to the work done by AYRS and, although still fairly primitive and cheap to make, they are effective and strong (Fig 67).

The basic elements of a horizontal-axis vane are, firstly, the lower bracket on which the whole outfit rests and on which the vane is rotated for course setting; secondly, rotating on the lower bracket, the bracket which carries the vane itself. On the vane is the line pulley which carries the steering line.

As the power of the vane is dependent on the leverage exerted by a big vane on a comparatively small pulley to which the steering lines are attached, a length of about four times the width makes a good vane. Two different sized vanes, one 1,200mm × 305mm and the other 1,500mm × 380mm, will cover most requirements.

The OGT Mark I

Fig 110 shows the construction details of a vane 1,500mm × 380mm with lines going direct to the tiller, designed to control boats up to about 9m, depending on how easy they are to steer. The vane size is about as large as it is practical to go so that, if this vane which is cheap and easy to make will not do the job, a more sophisticated gear will be required. Points to be noted when building this gear are:

For the vane, 4mm plywood is suitable and, although a little floppy, is quite strong enough. If thicker ply is used the vane is heavier, the counterweight is heavier and the whole vane, because of its weight, is less responsive to light breezes.

The backward slope of the vane is to bring the centre of pressure of the wind on the vane directly over the vane axis. This has the advantage of making the vane easy to turn for course-setting and,

at the same time, once the course has been set, the wind pressure does not tend to alter it. We have already seen that quite a number of commercial vanes have the vane axis at an angle to the horizontal of from 7° to 25°, and that this is reputed to give some feed-back. My tests indicated no obvious benefit from an inclined axis which needlessly complicated the linkage. Also, the angle used was not sufficient to allow the vane to weathercock if the course-setting control was released. If the vane was inclined far enough to make the vane weathercock, course-setting and holding the setting became more difficult.

The leading edge of the vane is stiffened by having 35mm × 18mm strips glued each side. These stiffeners are extended to carry the balance weight. The upper part of the stiffeners is tapered, again to save weight.

The lower edge of the 4mm-thick vane is stiffened by having a piece of 4mm ply glued on each side. This is to make the vane strong enough to carry the brackets on which it hinges. These brackets should preferably be made of good quality 9mm plywood, as the 8mm bolts which form the hinge-pins have a fair load on them and may wear thinner ply.

The pulley, also glued to the lower part of the vane, is made of three thicknesses of 6mm plywood glued together. By using three pieces it is easy to form the groove before the pieces are stuck together. In order to make sure of perfect alignment when the sections are glued, a small hole should be drilled at the centre of each disc and the three discs held in position by threading a nail through all three holes until the glue sets.

The upper bracket is made of 6mm plywood with the exception of the ends which carry the hinge pins which should, preferably, be slightly thicker. There is a groove round the periphery made the same way as the vane pulley. The groove carries the course-setting lines. On the lower side of the bracket there is a spigot which projects through the lower bracket. The two brackets are held together by a disc bolted to the spigot.

When making both the upper and lower brackets it is easier to make all the pieces and then screw them together with screws of about 18mm length. The parts can then easily be dismantled, glued and screwed together. This way, the whole unit can be glued up at one time. This saves having to wait for the piece which has just been glued to harden before the next piece can be added.

Centrally on the upper bracket, but inclined at a small angle to the vane axis, is a slot to carry a pulley and sheave such as is used for internal halliards on a dinghy mast. The reason for the small angle of inclination is to prevent the two tiller lines which are led over the pulley from chafing against each other.

The lower bracket, also made of 3 laminations of 6mm plywood, is glued by means of plywood gussets to a 46mm× 46mm square wooden post. This post is fixed to the boat to suit the layout. A bracket, as shown in the drawing, to hold it to the upper rail of the pushpit with the base of the post fixed to the deck, makes a good job.

The tiller lines are rove as shown in the drawing. The view is from aft looking forward with the balance-weight arm facing forward. It is quite easy to get these lines the wrong way round which gives rather disappointing results.

In use, the vane is rotated by the course-setting lines until it is vertical, with the stiffened edge facing into wind. The piece of chain which joins the two ends of the tiller lines is then dropped over the pin on the tiller.

The boat will probably luff up a little until the gear is at a sufficient angle to the wind to control the tiller. After this, adjustment of the angle of the vane by means of the tiller lines is simple. When not in use, the tiller lines should be cleated, as this prevents the vane rotating by itself.

The wooden post, if fixed to the push-pit, must be long enough to allow the counterbalance weight to clear the top rail. This means that the vane is quite a distance up in the air. This is ideal as there is no obstruction from people or things on the deck, also the wind strengthens very considerably a few feet above the water.

The balance weight should, initially, be heavy enough to bring the vane up to vertical, but not much more. If the steering on the run is erratic, increasing the balance weight has the effect of synthetic feed-back and can effect a quite remarkable improvement.

The tiller lines should be of stretchless polyester about 4mm or 5mm in diameter. Larger diameter lines are not required from strength considerations and only increase the friction over the pulleys. The pulleys themselves should be of 50mm diameter, as smaller sizes again increase friction.

The various parts of the gear have been described in considerable detail, basically to enable anyone with usual carpenter's tools to make one. If however someone wants to make the gear in aluminium, he can get all the basic ideas from the drawing. I know of one gear which has been made thus and which looks well and works well.

The pin on the tiller should be about 350mm to 450mm away from the rudder axis. If the vane flops from side to side, move the pin nearer the rudder axis which will give the vane more to do. If the vane cannot move the tiller enough, put the pin further away from the rudder; this will give the vane more purchase on the rudder. If there is not enough rudder movement to counteract weather helm, setting the tiller off centre by using a

Fig 110 (overleaf) *Self-steering vane OGT Mk 1;* (overleaf, opposite) *OGT Mk 1 details.*

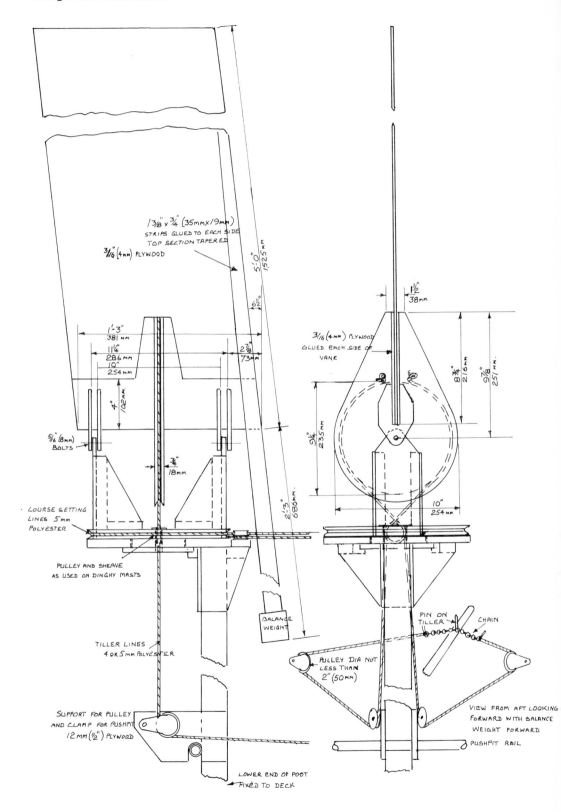

$1\frac{3}{8}'' \times \frac{3}{4}''$ (35mm x 19mm) STRIPS GLUED TO EACH SIDE TOP SECTION TAPERED

$\frac{3}{16}''$ (4mm) PLYWOOD

$5'-0''$ 1525 mm

$6\frac{1}{2}°$

$\frac{3}{16}''$ (4mm) PLYWOOD GLUED EACH SIDE OF VANE

$1\frac{1}{2}''$ 38mm

$8\frac{3}{4}''$ 216mm

$9\frac{7}{8}''$ 251mm

$9\frac{1}{4}''$ 235mm

$1'-3''$ 381 mm

$11\frac{1}{4}''$ 286 mm

$10''$ 254 mm

$2\frac{7}{8}''$ 73mm

$4''$ 102 mm

$\frac{5}{16}''$ (8mm) BOLTS

$\frac{3}{4}''$ 18mm

$10''$ 254 mm

COURSE SETTING LINES 5mm POLYESTER

$2'-3''$ 686 mm

PULLEY AND SHEAVE AS USED ON DINGHY MASTS

BALANCE WEIGHT

TILLER LINES 4 OR 5mm POLYESTER

PIN ON TILLER

CHAIN

PULLEY DIA NOT LESS THAN $2''$ (50mm)

SUPPORT FOR PULLEY AND CLAMP FOR PUSHPIT 12 mm ($\frac{1}{2}''$) PLYWOOD

VIEW FROM AFT LOOKING FORWARD WITH BALANCE WEIGHT FORWARD

PUSHPIT RAIL

LOWER END OF POST FIXED TO DECK

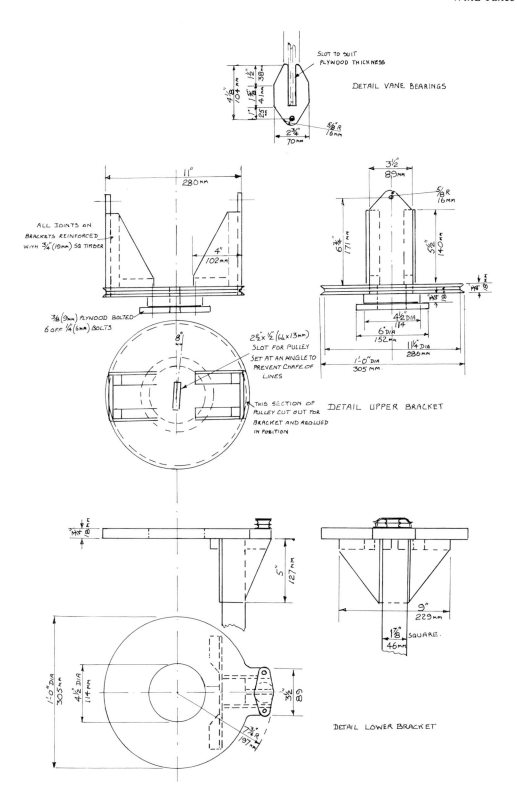

SLOT TO SUIT PLYWOOD THICKNESS

DETAIL VANE BEARINGS

ALL JOINTS ON BRACKETS REINFORCED WITH 3/4" (19mm) SQ TIMBER

4" 102mm

11" 280mm

3/8 (9mm) PLYWOOD BOLTED 6 OFF 1/4 (6mm) BOLTS

3 1/2" 89mm

5/8 R 16mm

6 3/4" 171mm

5 1/2" 140mm

1/4" 8mm

1/4" 8mm

4 1/2" DIA 114

6" DIA 152 mm

1 1/4" DIA 286mm

1'-0" DIA 305 mm.

2 1/2" x 1/2" (64x13mm) SLOT FOR PULLEY SET AT AN ANGLE TO PREVENT CHAFE OF LINES

8°

THIS SECTION OF PULLEY CUT OUT FOR BRACKET AND REGLUED IN POSITION

DETAIL UPPER BRACKET

1/4" 8mm

5" 127mm

9" 229mm

1 7/8" SQUARE. 46mm

1'-0" DIA 305 mm

4 1/2" DIA 114 mm

3 1/2" 89

3 7/8" R 97 mm

DETAIL LOWER BRACKET

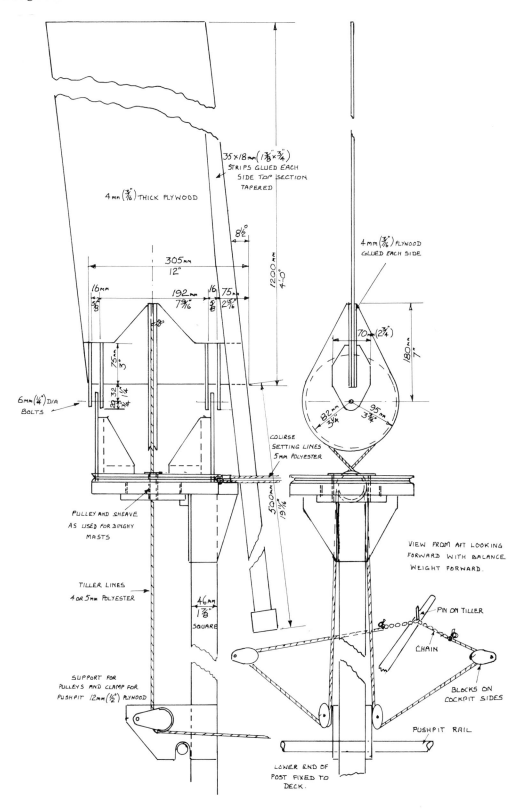

35×18mm (1⅜"×¾") STRIPS GLUED EACH SIDE TOP SECTION TAPERED

4mm (³⁄₁₆") THICK PLYWOOD

4mm (³⁄₁₆") PLYWOOD GLUED EACH SIDE

8½°

305mm
12"

1200mm
4'-0"

16mm
⅝

192mm
7⁹⁄₁₆"

16
⅝

75mm
2⁵⁄₁₆"

70mm (2¾")

180mm
7"

75mm
3"

32
1¼

82mm
3¼"

95mm
3¾"

6mm (¼") DIA BOLTS

COURSE SETTING LINES 5mm POLYESTER

PULLEY AND SHEAVE AS USED FOR DINGHY MASTS

500mm
19⁹⁄₁₆"

VIEW FROM AFT LOOKING FORWARD WITH BALANCE WEIGHT FORWARD.

TILLER LINES 4 OR 5mm POLYESTER

46mm
1⅞"
SQUARE

PIN ON TILLER

CHAIN

BLOCKS ON COCKPIT SIDES

SUPPORT FOR PULLEYS AND CLAMP FOR PUSHPIT 12mm (½") PLYWOOD

PUSHPIT RAIL

LOWER END OF POST FIXED TO DECK.

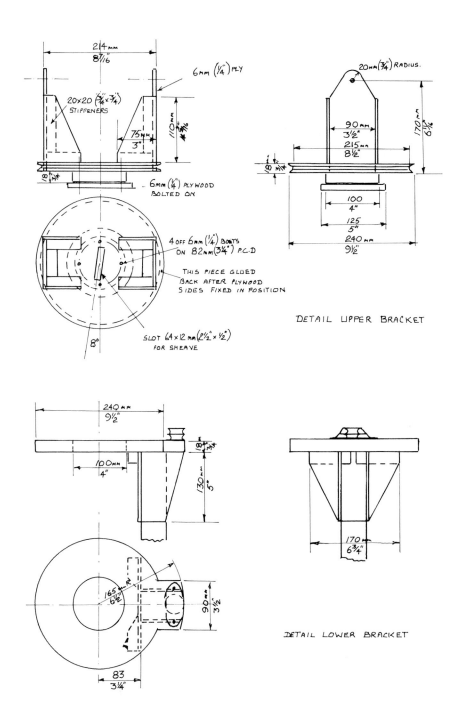

214mm
8 7/16"

6mm (1/4") PLY

20x20 (3/4" x 3/4") STIFFENERS

75mm
3"

110mm
4 3/8"

6mm (1/4") PLYWOOD BOLTED ON

18 3/4"

4 OFF 6mm (1/4") BOLTS ON 82mm (3 1/4") P.C.D

THIS PIECE GLUED BACK AFTER PLYWOOD SIDES FIXED IN POSITION

8°

SLOT 64 x 12 mm (2 1/2" x 1/2") FOR SHEAVE

20mm (3/4") RADIUS.

90 mm
3 1/2"

215mm
8 1/2"

170 mm
6 3/4"

18 3/4"

100
4"

125
5"

240 mm
9 1/2"

DETAIL UPPER BRACKET

240 mm
9 1/2"

18 3/4"

100mm
4"

130 mm
5"

165 1/2" R

90 mm
3 1/2"

83
3 1/4"

170
6 3/4"

DETAIL LOWER BRACKET

Fig 111 (opposite) *Self-steering vane OGT Mk 2;* (above) *OGT Mk 2 details.*

different link on the chain will cure this trouble.

A sample of the boats which I know to work well with this gear are:

A 14m Wharram catamaran (only just)
A 12m trimaran (*Piver Lodestar*)
Josephine, 10.7m overall with a balanced rudder
Hey Jude, 9.7m fin keel with a balanced rudder
A 9m sloop of modern design with a skeg-supported rudder
Tabitha, 7.5m with a poor design of rudder and a lot of hull-induced weather helm

Thus, the range of utility of this vane used directly on to the boat's tiller is from 7.5m to about 12m, dependent on the type of boat and the type of rudder.

The OGT Mark II

Fig 111 shows the basic parts of a smaller gear than the Mark I. This is designed for use either with boats below 7m with the lines coupled to the tiller or, when an auxiliary balanced rudder is fitted on the boat's stern, to control boats up to about 12m overall.

The whole arrangement is the same as the Mark I; the only difference is a smaller vane and mounting which, following our cube law, has about half the power of the larger vane. As the vane is not so heavy, the hinge-pins can be 6mm diameter only; the smaller the pin diameter, the less friction. The brackets have also been reduced in size. The result is a vane which is smaller in every respect which will not look so odd on a boat of, perhaps, 5m or 6m overall only.

Apart from being used to control an auxiliary rudder, the vane can be used direct coupled on most boats below 7.0m but, again, how well it behaves depends on the boat. As a guide, *Raha* — a Guy Thompson T24 — worked perfectly with one of these vanes, but *Raha* was well balanced and easy to steer. It certainly would not be man enough for the 7.5m *Tabitha*.

I have several times been asked 'What does OGT stand for?' In my student days anyone who went in for floating rather doubtful companies was accused of selling shares in the 'Ocean Going Tramway and Patent Back Collar Stud Corporation.' Back-collar studs are no longer with us, but there seemed to be no reason why 'Ocean Going Tramway' should not describe what was hoped to obtain by fitting a self-steering gear.

CHAPTER TWELVE

Balanced Rudders

Any rudder which has part of its blade forward of the rudder axis is easier to steer than an unbalanced rudder. And anyone who has sailed a dinghy with a kick-up rudder will know how the feel of the rudder changes if it is not in the fully down position. Ideally, we do not want to weaken the rudder stock too much by cutting too much away, so a shape with the centre of effort as far forward as possible is an advantage.

The NACA 00 series, which is a symmetrical airfoil developed in the USA, has several advantages as a rudder shape. It has a centre of effort which is practically constant at 24 per cent of the chord from the leading edge. Other sections have centres of pressure further aft and the position of these centres varies with the angle of incidence of the foil. It is also a good shape for strength, as the trailing edge does not taper to a fine section which would make it too liable to damage. The table on page 21 gives the offsets for the section based on a 10 per cent thickness, ie the width of 10 per cent of the chord; these offsets being given as a percentage of the chord. If the thickness is more or less than 10 per cent, the offsets can be modified to suit.

To construct a rudder based on this section, first of all decide on the chord to be used, then how thick the rudder has to be. This should be about 15 to 18 per cent for strength considerations. When the rudder is a plain balanced rudder a section tapering from top to bottom can be used, but when it is to be used with a trim tab it is easier to make it of one section all the way down.

Now draw the section full size, enclosing it in a rectangle which will be the size of the wood we start with. Then draw tangents to the section, the number of tangents depending on how accurate you wish the shape to be. If it is desired to make a tapered blade or rudder, a second shape is then drawn for the section at the bottom of the rudder.

Fig 112 is an actual design of a pendulum servo blade. First the four outside flats are carved or planed away, making certain that the flats are quite flat and are dead to the lines marked on the blank. The next flats to be cut are then marked out and planed away. Then the whole section is trued up by feel, making certain that it is only the corners which are planed away and not the parts of the flats which are tangents to our shape.

The rudder has now to be carved away for the pintles (Fig 113); but as this is weakening the rudder stock, care should be taken to make pintles which only require the minimum of cutaway. The centre of the pintles should be approximately 23 per cent of the chord back from the leading edge. This must be made as accurate as possible, as at 25 per cent the rudder will probably be unstable and at less than 20 per cent the vane is being called upon to do more work than necessary. Note that any imperfection in the making of the rudder will move the centre of pressure from the theoretical 24 per cent position to further aft. The centre of pressure will never be at less than 24 per cent of the chord.

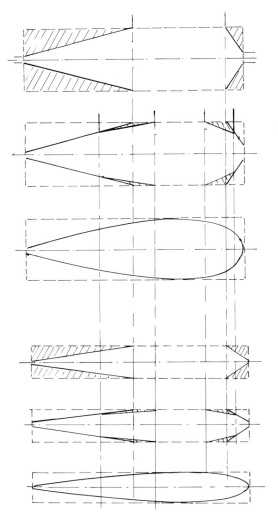

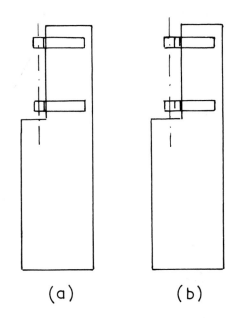

Fig 113 (a) Good and (b) bad pintle design. The
pintles should be made so that there is the
minimum cut out from the rudder stock.

Fig 112 Construction of pendulum-servo blade
by means of tangents.

The size of the balanced rudder when
used as an auxiliary rudder should be
about half that of the main rudder. As a
rough guide, up to 9m a 300mm chord ×
1,000mm immersion will control the boat
with some assistance from the main
rudder. Boats of up to 12m will need a
rudder of 380mm × 1,100mm. Above this
size of boat, the rudder should be scaled
up accordingly. *Josephine*, of 10m overall,
used a main balanced rudder of 450mm ×
1,100mm, but this was her only rudder
and so was larger than if it had been an
auxiliary. The boat was designed to have
the minimum of weather helm and the
rudder was, if anything, oversized.

The best material for making a
balanced rudder is wood. This can be
either a single piece carved to shape if the
timber is not liable to crack or, for greater
security, several pieces of thin wood
laminated together using a good quality
marine glue. It is not recommended that a
rudder should be fibreglassed. With pro-
longed immersion some water will nearly
always find its way into the wood due
either to porosity or slight damage. The
wood will then swell and rupture the
fibreglass. This looks bad. If the rudder is
painted with a good quality paint and
then antifouled, there should be no
trouble.

Weighting the Rudder

It has been mentioned earlier that before
the invention of the Braine gear in 1902,
model yachts used a weight on the rudder
so that when the boat heeled the weight

applied rudder, which checked the weather helm. The balanced rudder, if weighted, will have the same effect and on the run, when the wind is light and the boat rolling, a weighted rudder will help to keep the boat on a straight course without any assistance from the wind vane. The optimum weight depends on the exact shape and buoyancy of the rudder and on the shape of the boat. Some boats are directionally far more stable than others. Trying different weights is not difficult and is well worth the time spent, particularly if the boat is to go trade-wind sailing with the wind from astern for days on end.

Now that we have got our balanced rudder the next question is how do we control it with a wind vane. A vertical-axis vane will work a balanced rudder, but its power is really not suited for the job. The horizontal-axis vane has plenty of power and it is this type that is recommended.

Sometimes the transom of the boat slopes downwards. The balanced rudder fitted to *Nemesis* was on a 22½° slope and gave no trouble (Fig 114). How far rudders could be inclined is open to speculation, but it is worth experimenting as a rudder fitted directly to the transom is a lot neater than having to make a complicated framework to enable the rudder to hang vertically.

Practical Balanced Rudders

The OGT Mark II (Fig 111) was designed to control a balanced auxiliary rudder by means of tiller lines. By this method a cheap and easily constructed gear can be made for normal boats up to about 12.0m overall. The plain balanced rudder may need more vane force than a trim-tab assembly but, as has been shown by the test results, gives a better performance as a rudder, as there is no trim tab working against it. The plain balanced rudder is considerably less complicated to make

than the rudder and trim-tab assembly and is recommended where speed of making, or cost, is the first consideration.

There are two rudders shown. One rudder was designed so that we could cruise *Nemesis*, an 11.5m sloop, back from the Noumea race and get some time away from the tiller to let the crew have instruction in navigation. The gear did all that was required of it, but I do remember on one occasion the skipper being more than somewhat annoyed as all his crew were down below playing cards, leaving the boat to look after itself. The rudder is at an angle and not vertical. I designed it like this so that, apart from fitting pintles to the boat and carrying the rudder and vane, there was no major obstruction which could interfere with the race. The angle did not seem to affect the steering adversely. Subsequently this rudder and vane were used on an 11m ferro-cement sloop which had to be delivered at short notice to Rarotonga. It worked well until it fell off, as the boatyard which fitted the pintles had failed to grout the bolts in position. It was repaired and brought the same boat back to Auckland without trouble.

The second rudder has a more extensive mounting to allow it to hang vertically and was used on the 11.5m ketch *Portofino* on a delivery trip to Suva. This gave us no trouble in spite of having a duplicated vane system so that the boat could sail with the mizzen hoisted. The extra complication of the mounting was justified as the boat was to cruise in the Fijian Islands.

In each case the section of the rudder has been shown as an example of how to mark out a blank before shaping it. The order of the removal of the unwanted material is given: A is the first cut, then B, C etc. It will be seen that there appears little relationship between the two rudders; they were made at different times and without reference to each other. If the shape of the rudder is drawn out

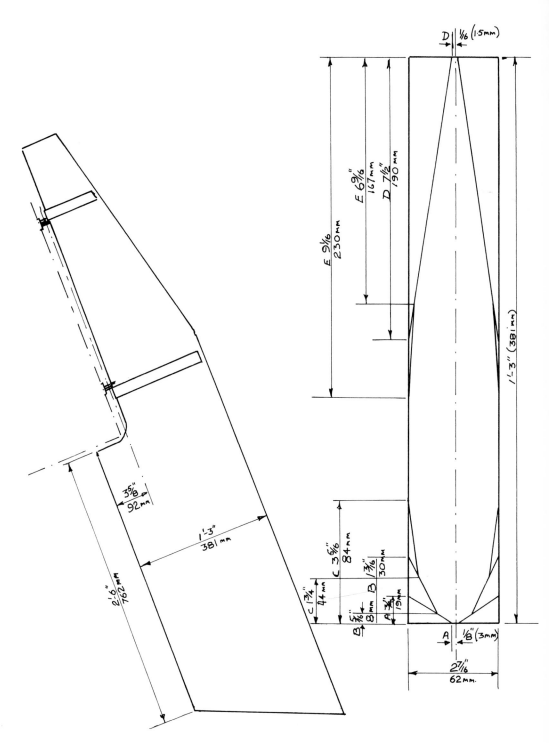

Fig 114 Auxiliary balanced rudder for
Nemesis.

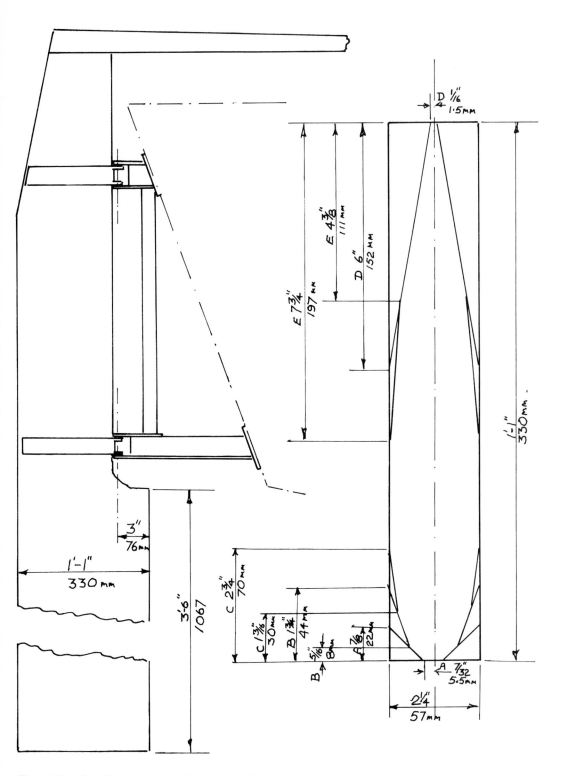

Fig 115 Auxiliary balanced rudder for
Portofino.

using the table of offsets it does not matter how the tangents for the cuts are drawn as long as the result gives the shape we want. In any case, both rudders worked. The *Nemesis* rudder was laminated from three layers of nominal 1in planed timber. This timber was $1\frac{1}{16}$in thick and gave an overall thickness of $2\frac{7}{16}$ in, or 62mm. When the *Portofino* rudder was made, New Zealand had metricated and the timber was 19mm, giving 57mm overall. In each case the balance of the rudder was between 22 per cent and 23 per cent, whilst the thickness ratio was 16 per cent to 17 per cent.

The rudders were not fibreglassed as it needs only a very small graze to allow water to penetrate to the wood which then swells and ruptures the glass. This looks bad. It is sufficient to paint the wood with good quality paint and then antifoul. The timber used does not seem to be important; both these rudders were made from tanalized pine which is very resistant to all forms of marine borer even if the paint film is damaged. As regards glue, any form of good quality marine type glue will do the job. Having flown in plywood aeroplanes during World War II, I have my own personal preference for Aerolite, possibly due to the fact that I am still alive. Any of the glues on the market are probably just as good provided they meet marine specifications.

The metal work of the pintles can be either stainless steel or mild steel galvanized, or, if the gear is for a delivery trip only, plain mild steel painted. Of the two examples, one was designed for a single trip only and painted steel was quite good enough. *Portofino*, whose gear was to be a permanent feature of the boat, had all metalwork in stainless steel.

The tiller attached to the rudder should be about 450mm long as this allows enough length for the control-line attachment and, when manoeuvring in restricted water, an extra rudder right at the end of the boat does wonders to the turning circle, but only if there is a tiller and an extra hand to work it.

CHAPTER THIRTEEN

Auxiliary Rudder with Trim Tab

Before the invention of pendulum servo gears nearly all designers used trim tabs as a means of stepping up the force produced by the vane so that it could operate a rudder. Today trim tabs are normally used only for boats with rudders projecting over the stern which get in the way of any other gear, and where the fitting of a trim tab presents no difficulty. There are, however, many commercial gears made in the form of an auxiliary balanced rudder and trim-tab combination worked either by a vertical or horizontal axis vane. These are easy to fit onto the boat and eliminate all the complications if the boat is wheel steered, or has a central cockpit. The independent rudder needs only one control for course-setting. This is easily led to the steering position and is far less complicated than the several lines required for a pendulum servo gear.

Fig 116 shows a complete gear consisting of balanced rudder, trim tab and vertical-axis vane all mounted on its own frame ready to fit on the stern of a boat. This gear is a far more sophisticated outfit than the plywood flop-over vanes and requires stainless steel welding. It is a job for a well equipped workshop rather than the backyard garage. If a gear of this type is wanted and only simple tools are available, a new design will have to be made using simpler materials, but keeping the measurements basically the same.

The vane is a standard vertical-axis vane as described on page 36. The course-setting is by means of a drilled disc and pin, controlled by a line from the steering position. To set a course the pin is disengaged by pulling the course-setting line. The boat is then steered onto a course slightly below that which is wanted and the wheel or tiller fixed in position. The best way to do this is with lines which are adjusted by means of the wooden blocks used to tension tent ropes (Fig 117). When the boat is left to itself it will luff up slightly until there is enough force of the wind on the vane to work the trim tab, when the self-steering will take over.

As regards the making of the drilled disc, it is obvious that the holes must be fairly accurately drilled or else the pin will not fall into position. It is better to fix the vane shaft to the drill base and then put the disc on to it (Fig 118). A hole is drilled and then the disc rotated for the next hole, but the distance from the vane shaft to the drill does not alter.

As Fig 116 shows, the linkage between the course-setting and the trim tab gives a reasonable degree of feed-back, as the vane shaft is on the trim-tab axis with the distance from the trim-tab axis to the rudder axis 268mm and the length of the vane-lever arm 216mm. This gives a ratio near enough to the 80 per cent recommended.

The rudder and trim tab are marked out in exactly the same way as a plain rudder. The trim tab is 95mm wide and the overall width of rudder and trim tab 445mm, which gives a ratio of tab to overall of 21.5 per cent which is slightly more than our recommended 20 per cent, but

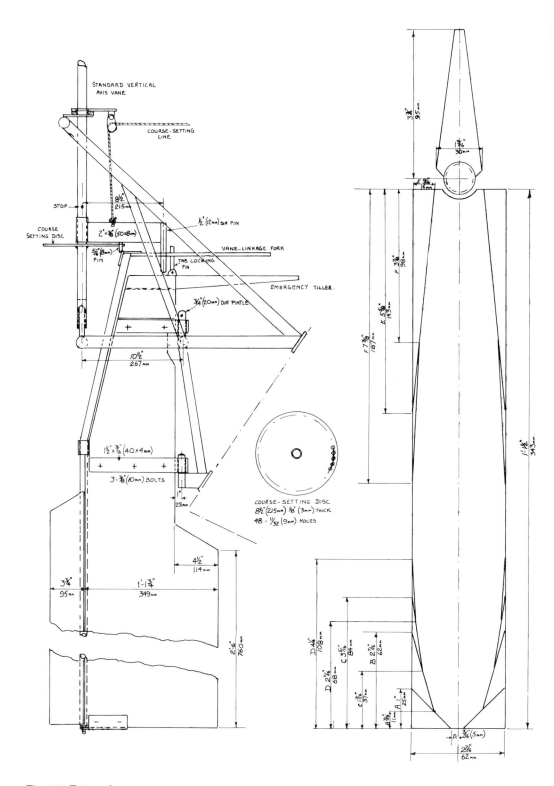

Fig 116 Trim-tab gear.

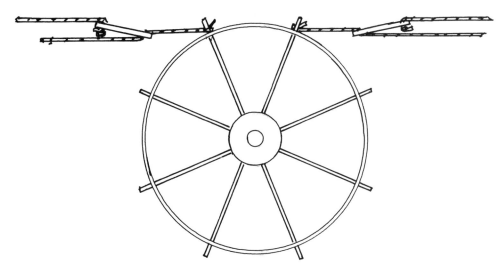

Fig 117 Tent guys are useful to fix the boat's wheel or tiller when an auxiliary rudder is in use.

not so far away as to spoil the gear. The thickness of the rudder is 62mm, or 14 per cent. The amount of balance, that is to say the distance from the rudder axis to the leading edge, is 90mm, giving a balance of 20 per cent which is, perhaps, a little conservative and could be increased to about 23 per cent, probably with improved performance. The framework on which the gear is mounted is made from 25mm and 30mm stainless steel pipe. The pads which are bolted to the boat have, of course, to be made to fit the individual boat.

One detail which should not be over-looked is that the rudder must be very free on its pintles so that it can respond to each little variation of the trim tab. If the pintles are separately bolted to the boat it is very hard to get them dead in line; therefore it is always advisable to have the pintles set accurately on the framework and then bolt the framework to the boat. By this means, any lack of fairness of the boat does not upset the line of the pintles.

A tiller is mounted on the rudder to control the boat, particularly when reversing in marinas, and for this purpose

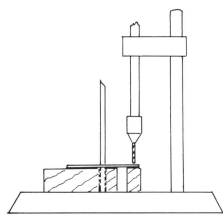

Fig 118 If a course-setting disc is set up in the drill press, all the holes are certain to be on the exact radius.

there is also a latch so that the trim tab can be fixed central and the auxiliary rudder and tab form one unit.

The size of a rudder without its trim tab should be slightly larger than the size required if a balanced rudder only is used. This is because the trim acts against the rudder. At small angles of deflection the rudder and tab act as a single foil and the rudder is as effective as a single balanced rudder. Under extreme conditions, when the self-steering is hard pressed, the efficiency of the trim-tab gear falls off because the water flow over the trim-tab combination is very

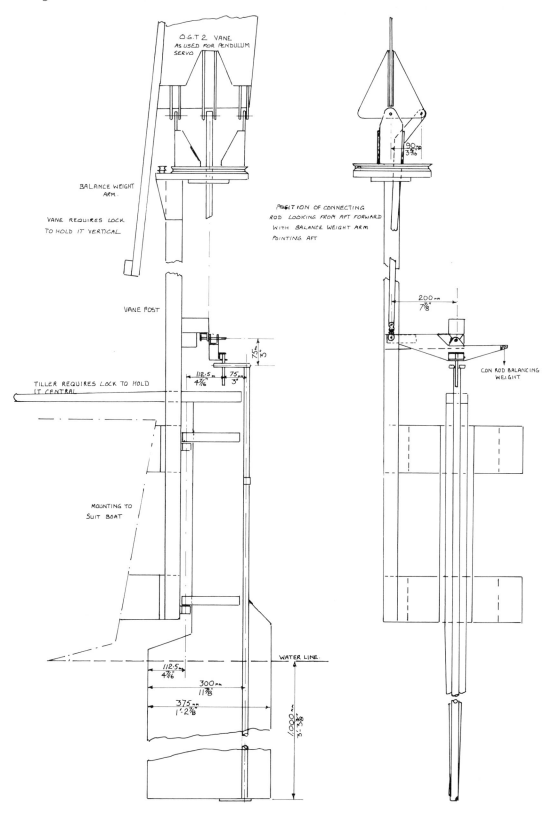

O.G.T 2 VANE
AS USED FOR PENDULUM
SERVO

BALANCE WEIGHT
ARM.

VANE REQUIRES LOCK
TO HOLD IT VERTICAL.

POSITION OF CONNECTING
ROD LOOKING FROM AFT FORWARD
WITH BALANCE WEIGHT ARM
POINTING AFT

90 mm
3 9/16"

VANE POST

200 mm
7 7/8"

75 mm
3"

112.5 mm
4 7/16"   75 mm
           3"

TILLER REQUIRES LOCK TO HOLD
IT CENTRAL

CON ROD BALANCING
WEIGHT

MOUNTING TO
SUIT BOAT

WATER LINE.

112.5 mm
4 7/16"

300 mm
11 7/8"

375 mm
1'-2 7/8"

1,000 mm
3'-3 3/8"

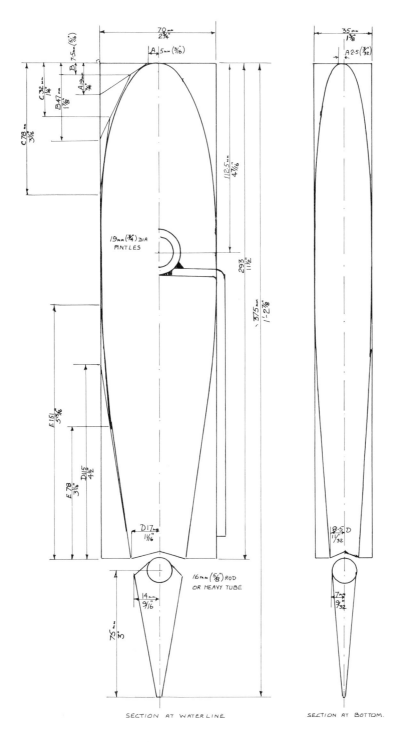

Fig 119 (opposite) *Autoptère self-steering gear;* (above) *Autoptère rudder sections.*

irregular, and it stalls out before an equivalent balanced rudder would develop its full potential.

I have used the following sizes with success and boats using these sizes have all made ocean passages:

Boat Length m	Vane mm	Rudder Chord mm	Trim Tab mm	Immersion in Water mm	Rudder Thickness mm
10	standard vane 600 × 1,250	330	82.5	760	63
12	standard vane 600 × 1,250	380	95	760	76
18	680 × 1,830	455	102	760	83

Note: The depth of immersion is that at rest. Usually this is considerably exceeded. In all cases the rudders were made with 20 per cent to 23 per cent balance.

This gear is not difficult to make and seems to be pretty foolproof, the single control line being a great advantage in a crowded cockpit. Care should be taken that the mounting onto the boat is very sturdy as not only is there a tendency for the framework to be used as a boarding ladder but also, in extreme conditions such as a boat coming sideways off a wave top, very considerable stresses are set up in the rudder mounting and rudder stock. This is not as extreme as it might be because the main rudder, which is fixed, takes a lot of the yawing force; also because, as the angle of the rudder to the water flow passes about 20°, the centre of pressure moves aft and the whole rudder and trim tab will be turned against the wind vane rather than try to resist the motion.

Autoptère Rudder

The Autoptère rudder (Fig 119) is not easy either to design or to make up. In designing, the following points have to be considered: the size of the rudder, the size of the trim tab, the amount of over-balance of the pintles, the feed-back ratio,

and the linkage ratio between vane and trim tab. Nearly all these factors are interdependent — by changing one another is affected.

The overall size of the rudder can be taken as the same as a plain balanced rudder doing the work required. In actual fact the Autoptère rudder is probably slightly more efficient than a symmetrical rudder as it is curved the right way, but this difference is not great.

The size of the trim tab is dependent on the out of balance of the rudder pintles. Our test rudder had a 20 per cent tab with 31 per cent balance of the rudder. This combination needed about 15° of trim-tab deflection to give 12.5° of rudder. With the sizes used this means that with 12.5° deflection of the rudder the trim tab was at an angle of 11° relative to the rudder, which is near enough to our design requirement that the deflection of the trim tab relative to the rudder should be the same as the deflection of the rudder relative to the boat. In view of this we stuck to the 20 per cent trim tab with a maximum deflection relative to the boat of 20° which gives us about 15° of rudder, which is probably its maximum effective angle.

As we were using a 20 per cent tab the balance of the rudder was kept at 30 per cent. This is not exactly the same as our test rudder, but to use 31 per cent seemed a little pedantic.

The feed-back ratio was kept at 40 per cent, as this seemed to work and was a nice round number.

The linkage ratio required a tab movement of 20°. A horizontal-axis vane required a ratio of 45:20 if the vane was to swing 45° each side of the vertical. Using a vertical-axis vane a 1:1 ratio would require a 20° vane movement to give maximum rudder deflection. This would result in sloppy steering and so the ratio used was 1.5:1 of tab movement to vane movement.

The initial tests of the completed

rudder were made with a vertical-axis vane, as one with a course-setting disc happened to be available. This gear demonstrated the advantage of the Autoptère rudder which was much more powerful than the normal equivalent trim-tab set-up. Control was positive but did not, on the boat used, give oversteer, and there seemed to be no undue wandering on the run. There were, however, major snags. In order to change course with a pin and disc course-setting, the pin must be released from the disc. As soon as this happens with an Autoptère, the rudder goes out of control as there is no longer the restraining influence of the trim tab. To change course either the rudder had to be held central, in which case the trim tab would centralize itself, or the trim-tab pin had to be fixed centrally so that it controlled the rudder.

In any case, course-setting was obviously going to be complicated if it was to be done from a distance. Thus the vertical-axis vane was scrapped and an OGT Mark II horizontal-axis vane substituted using the same linkage as for the pendulum servo. This cured the course-setting difficulty as the tab pin was never free. When hand-steering was required the tab pin could be held central by fixing the vane in the vertical position. This would enable the boat to be steered either by the tiller on the Autoptère, or the boat's main rudder could be used with the auxiliary left to trail. Everything seemed fine until the motor was started. Then the Autoptère went haywire, as the wash from the propeller hit it on one side and the whole steering went out of control. The only way to regain control was to fix the Autoptère tiller central and steer the boat with the main rudder. There was a further incident when manoeuvring with the Autoptère swinging over too far and disconnecting the fork and pin linkage.

The moral of this chapter of accidents is that the Autoptère is best controlled by a horizontal-axis vane. The OGT Mark II is probably a little oversized, but this is no defect to the working of the gear as such a vane is very positive in its control of the trim tab. The linkage should have a ratio of 2:1. There should be a catch to hold the vane vertical for hand-steering under sail. There should be a means of holding the Autoptère tiller central when under motor and there should be stops to prevent the Autoptère rudder swinging more than, perhaps, 25° each way. With these alterations the gear gave good results.

Marcel Gianoli found that he had to use two axes on his MNOP 66 gear, one for the overbalanced rudder and another when the rudder and trim tab were being used manually. This was obviously a more expensive arrangement and one wonders where the need for two axes arises. The Mustafa gear, which also uses an Autoptère blade, has only one axis for the rudder, so whether the gear works this way or not is perhaps a question of design and linkage.

CHAPTER FOURTEEN

Pendulum Servo Gear

I have designed several types of pendulum servo gears, the already mentioned one of all wood construction for home manufacture for the New Zealand magazine *Sea Spray* and, at the other end of the scale, a gear of stainless steel for commercial manufacture. The gear described here has been developed to use a horizontal-axis vane.

The vane is a standard OGT Mark II (Fig 111). This is, perhaps, a little oversize for the work it has to do, but should give a very positive result. Tests on the complete gear indicated that about 300mm could be cut off the top of the vane without appreciable difference, depending on the user's preference as to whether he wants a large or small vane. The vane, in place of the pulley for the lines, is fitted with a mounting for the connecting rod. The top bracket, in place of the slot 64mm × 12mm for the line pulley, has a larger slot 70mm × 30mm at right angles to the vane axis and not inclined as the line pulley is. This slot is for the connecting rod which works the pendulum servo blade.

The connecting rod can be either of light metal pipe or of wood. It has a crank at the top so that it passes through the slot in the top bracket. The lower end has a swivel, so that the vane can be turned round for course-setting without affecting the blade lever arm. The connecting-rod swivel must be as near as possible exactly on the course-setting axis of the vane.

The height of the 46mm × 46mm post which supports the wind vane should be

long enough for the balance weight of the vane to clear the pushpit. This measurement should be checked and, if necessary, the connecting rod and post lengthened to suit.

There are stops on the main framework so that the lever arm can only rotate 15° each side of the centre line. This also limits the travel of the vane to 45°. The blade lever arm has a small counter balance weight to exactly balance the weight of the connecting rod.

With any pendulum servo gear there must be some form of feed-back so that when the blade is at the end of its travel it automatically turns until it is facing direct into the water flow, thus having all strain taken off it. Without this the blade may swing over under full power until it can go no further due to the framework or other obstruction, and then excessive loading may break it. There are two methods of providing feed-back, already described on pages 53–4. This particular design uses geometrical feed-back where, as the blade swings over, it gradually straightens itself out. If the blade swings over too far it will actually be turned further than facing the water flow and will tend to return the oar towards the central position.

Fig 121 shows the geometry of the feed-back. The oar itself uses standard dinghy type pintles and is cut away to give 22 per cent balance. The position of the fork for the lever-arm pin is determined by the

Fig 120 (opposite) *Pendulum servo gear with* (overleaf) *its details.*

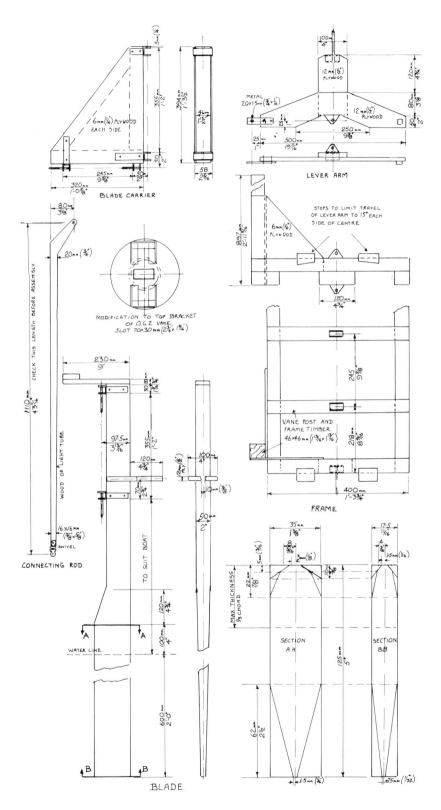

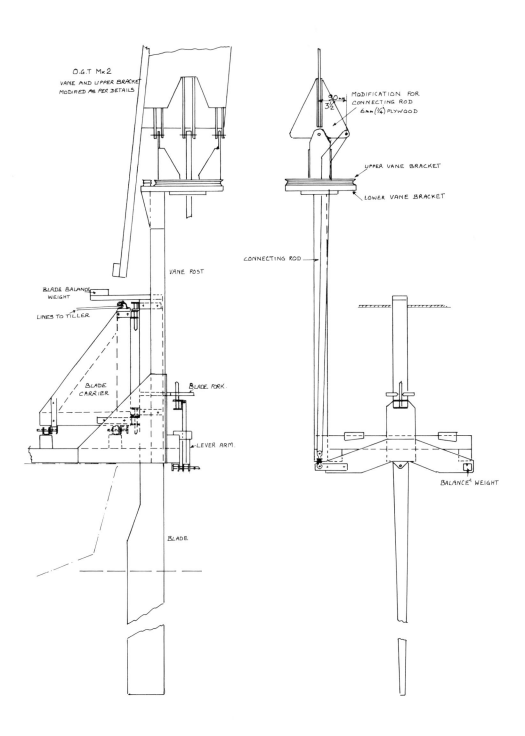

O.G.T Mk 2
VANE AND UPPER BRACKET
MODIFIED AS PER DETAILS

MODIFICATION FOR
CONNECTING ROD
6mm (1/4) PLYWOOD

30mg
3 1/2

UPPER VANE BRACKET

LOWER VANE BRACKET

VANE POST

CONNECTING ROD

BLADE BALANCE
WEIGHT

LINES TO TILLER

BLADE
CARRIER

BLADE FORK

LEVER ARM

BALANCE WEIGHT

BLADE

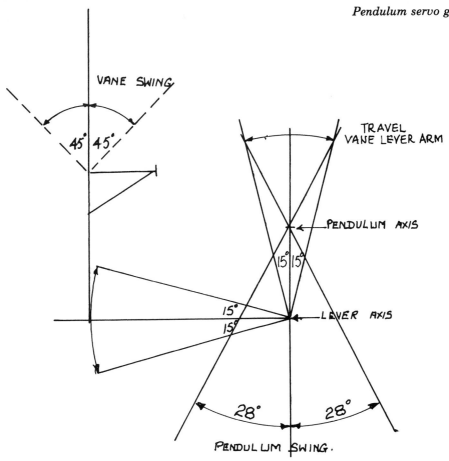

VANE SWING

45° 45°

TRAVEL
VANE LEVER ARM

PENDULUM AXIS

15° 15°

15°

LEVER AXIS

15°

28° 28°

PENDULUM SWING.

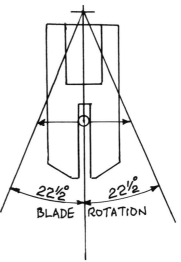

22½° 22½°

BLADE ROTATION

Fig 121 Geometry of the pendulum servo gear. This shows that the vane moves 45° each side of vertical. The pendulum swings 28° before feed-back eliminates the force on the blade. The blade is initially turned 22½° by the lever system. This seems excessive but as the blade swings over the feed-back will reduce the angle. This gives more power over the whole swing than would be the case if the blade was only deflected a maximum of, say, 17°.

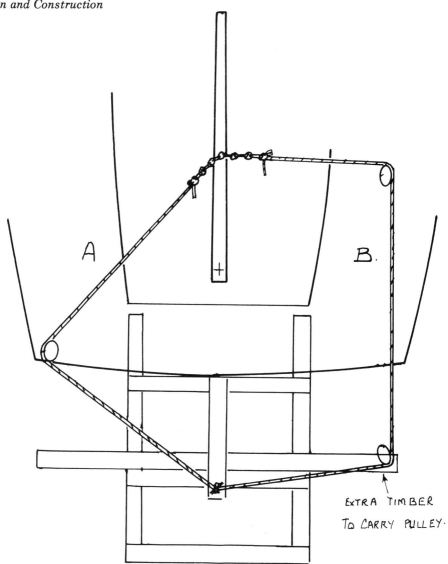

EXTRA TIMBER
TO CARRY PULLEY.

Fig 122 The method of leading the steering lines from the oar carrier to the tiller is determined by the shape of the boat: method A worked on Tabitha; *method B is probably better but requires extra framework to carry the pulleys.*

feed-back geometry. This blade cannot be hinged up as this introduces an extra complication and, in practice, I have found that a servo blade does not seem to get hit by debris, although sometimes seaweed gets caught in it. This can easily be removed with a boathook. In harbour, a blade hinged out of the water is very

subject to damage by visitors' dinghies and, if it is desired to take it out for any reason, it is better to remove it entirely. There is no difficulty in lifting this blade off its pintles.

At the top of the blade is a balance weight. The reason for this is that the bulk of the blade is behind the pintles and, with the blade swung over to one side and thus out of the vertical, the weight of the blade tends to turn it, giving positive feed-back. A wooden blade does not suffer very much from this as a lot of the blade is in the water and the buoyancy counteracts the weight. If however the blade is heavy, as when made of solid fibreglass, the problem becomes serious. Exactly weighting the blade about its pintles gives a form of synthetic feed-back along the lines of the weighted rudder. This weighting gives vastly better results in light winds, particularly on the run. The shape of the blade is based on the NACA 00 series. Although the material of the blade can be either wood or fibreglass, for 'one off' gears wood is perfectly adequate.

The blade carrier is of sandwich construction with 39mm × 46mm square timber inside and 6mm plywood on each side. The carrier is fitted with dinghy type pintles for a 50mm thick rudder. The dinghy pintles shown on the drawing are those available in Australasia; other countries may use a different type necessitating a slight alteration to the design to make them suit. The framework is of 46mm-square section timber and can be extended as required to suit the boat.

The tiller lines are fixed to the top of the blade carrier and are led from there over pulley blocks to the tiller. The exact lead depends on the shape of the boat; two alternative methods are shown in Fig 122.

To use the gear the tiller lines are fastened to the tiller-line chain so that, with the tiller central and the self-steering blade vertical, the central link of the chain fits over the pin on the tiller.

The tiller lines should not be too thick; about 5mm diameter is plenty as thick lines create extra friction as they travel over the pulley blocks. Initially there will be some stretch but, using polyester lines, this will disappear as the lines are worked in. The lines should not be too tight as this increases friction. The course-setting lines can be 3mm or 4mm lines led to a cleat in any convenient position.

To use the gear the wind vane is turned until it faces the wind and then the central link of the chain dropped over the pin on the tiller. This pin should be at about 375mm along the tiller from the rudder axis. If, under extreme conditions, it is found that the blade travels over to one side without giving enough tiller movement, then the chain can be offset on the pin. This has the effect of giving more tiller movement without so much blade movement. In an emergency it is necessary to disconnect the tiller lines as it is impossible to hand turn the tiller against the power of the gear. Disconnection is simply a matter of lifting the chain off the tiller pin.

In harbour, the blade can easily be lifted off its pintles if it is not going to be used for an appreciable time. It is better to prevent the blade from accidentally being lifted off its pintles by drilling one of the pintles and using a split pin. In a very rough and choppy sea I once had the blade come out of its pintles, which could have been very embarrassing.

It is hoped that this book has helped you to make your own gear, or to get better service from one you already have. I can now only wish you fine weather and good sailing, but please keep a lookout.

Bibliography

Aero-hydrodynamics of Sailing, Marchaj, C. A. (Adlard Coles, 1979)

Mechanics of Flight, Kermode, A. C. (Pitman, 1972)

Model Racing Yachts, Browne, D. (Rolls House Publishing Co, 1949)

Pen Duick, Tabarly, Eric (Adlard Coles, 1971)

Rudder Design for Sailing Yachts, Amateur Yacht Research Society (Hythe, Kent, England, 1974)

Sailing Theory and Practice, Marchaj, C. A. (Adlard Coles, 1964)

Self-Steering, Amateur Yacht Research Society (Hythe, Kent, England, 1974)

Self-Steering for Sailing Craft, Letcher, J. S. (Intnl Marine Pub Co, 1974)

Self-Steering for Yachts, Dykstra, G. (Nautical Pub Co, 1979)

National Advisory Committee for Aeronautics Technical Report 938, 'Summary of Section Data on Trailing-Edge High-Lift Devices' (US Government Publication, 1951)